Lecture Notes in Intelligent Transportation and Infrastructure

Series Editor

Janusz Kacprzyk, Systems Research Institute, Polish Academy of Sciences, Warsaw, Poland

The series "Lecture Notes in Intelligent Transportation and Infrastructure" (LNITI) publishes new developments and advances in the various areas of intelligent transportation and infrastructure. The intent is to cover the theory, applications, and perspectives on the state-of-the-art and future developments relevant to topics such as intelligent transportation systems, smart mobility, urban logistics, smart grids, critical infrastructure, smart architecture, smart citizens, intelligent governance, smart architecture and construction design, as well as green and sustainable urban structures. The series contains monographs, conference proceedings, edited volumes, lecture notes and textbooks. Of particular value to both the contributors and the readership are the short publication timeframe and the world-wide distribution, which enable wide and rapid dissemination of high-quality research output.

More information about this series at http://www.springer.com/series/15991

Amin Mobasheri
Editor

Open Source Geospatial Science for Urban Studies

The Value of Open Geospatial Data

Springer

Editor
Amin Mobasheri
GIScience Research Group
Heidelberg University
Heidelberg, Germany

ISSN 2523-3440 ISSN 2523-3459 (electronic)
Lecture Notes in Intelligent Transportation and Infrastructure
ISBN 978-3-030-58231-9 ISBN 978-3-030-58232-6 (eBook)
https://doi.org/10.1007/978-3-030-58232-6

This Springer imprint is published by the registered company Springer Nature Switzerland AG
The registered company address is: Gewerbestrasse 11, 6330 Cham, Switzerland

This book is dedicated to:

National liberation and those who have fought for it.

Ignorance warriors. As they are the key people ensuring equality.

Memories of those who have worked hard for us to have a better life now.

Mankind equal rights continuators.

Idealists. Since they act towards pervading global peace.

Preface

Nowadays, governments from around the world and stakeholders from the business sector both participate to and promote open geospatial science. Governments increasingly provide free access to various types of geospatial data as they realize its potential to foster economic, social, urban and environmental opportunities. Concrete projects based on open geospatial data are now having significant and measurable impact on communities, economy, environment, health and transportation, only to name a few areas. In this book, we focus on the benefits that open geospatial science in general, and open geospatial data, in particular, bring to urban studies with particular focus on transportation and smart city analytics projects. The book does not aim to cover all possibilities and potentials, since this is an impossible task to handle, but tries to include up-to-date practical studies that address some concrete challenges within the proposed domain. This book is suggested to be read by students and researchers of urban and transportation sciences as well as Geo-Information Science.

Heidelberg, Germany — Amin Mobasheri

Contents

Contributors

Berk Anbaroğlu Department of Geomatics Engineering, Hacettepe University, Ankara, Turkey

Anahid Basiri Centre for Advanced Spatial Analysis, University College London, London, UK

Huanfa Chen Centre for Advanced Spatial Analysis, University College London, London, UK

Sagi Dalyot Mapping and Geo-Information Engineering, Technion, Haifa, Israel

Tiago H. Moreira de Oliveira NOVA IMS, Universidade NOVA de Lisboa, Lisbon, Portugal

Yerach Doytsher Mapping and Geo-Information Engineering, Technion, Haifa, Israel

Talia Dror Mapping and Geo-Information Engineering, Technion, Haifa, Israel

Dogus Guler Department of Geomatics Engineering, Istanbul Technical University, Istanbul, Turkey

Ayse Giz Gulnerman Department of Geomatics Engineering, Faculty of Civil Engineering, Istanbul Technical University, Istanbul, Turkey

Himmet Karaman Department of Geomatics Engineering, Faculty of Civil Engineering, Istanbul Technical University, Istanbul, Turkey

Amin Mobasheri GIScience Research Group, Heidelberg University, Heidelberg, Germany

Alan T. Murray Department of Geography, University of California at Santa Barbara, Santa Barbara, CA, USA

Syed Monjur Murshed European Institute for Energy Research (EIFER), Karlsruhe, Germany

Sebastián Ortega CTIM, Universidad de Las Palmas de Gran Canaria, Las Palmas, Spain

Marco Painho NOVA IMS, Universidade NOVA de Lisboa, Lisbon, Portugal

José Miguel Santana CTIM, Universidad de Las Palmas de Gran Canaria, Las Palmas, Spain

Agustín Trujillo CTIM, Universidad de Las Palmas de Gran Canaria, Las Palmas, Spain

Jochen Wendel European Institute for Energy Research (EIFER), Karlsruhe, Germany

Tahsin Yomralioglu Department of Geomatics Engineering, Istanbul Technical University, Istanbul, Turkey

An Introduction to Open Source Geospatial Science for Urban Studies

Amin Mobasheri

Abstract Nowadays, governments from around the world and stakeholders from the business sector both participate to and promote open geospatial science. Governments increasingly provide free access to various types of geospatial data as they realize its potential to foster economic, social, urban and environmental opportunities. Concrete projects based on open geospatial data are now having significant and measurable impact on communities, economy, environment, health, and transportation, only to name a few areas. Hereby, we focus on the benefits that open geospatial science in general, and open geospatial data in particular bring to urban studies with particular focus on transportation and smart city analytics projects. This chapter introduces up-to-date practical studies that address some concrete challenges within the proposed domain, and ends with some remarks on the topic.

Keywords Open geospatial data · Open source GIS · Urban studies · Smart cities

1 Introduction

The advances in open source geospatial science has influenced the emergence of various applications that were restricted in former times. Geospatial technologies have nowadays become ubiquitous. There are various infrastructures that aim to provide services to collect, edit and freely share geospatial information. Examples of such infrastructures include OpenStreetMap, Flickr, Twitter, Foursquare for the usage of public volunteers. OpenStreetMap itself has gained severe interest and attention for research and applied purposes. The past decade has shown hundreds of projects implemented/running either on focusing on this Volunteered Geographic Information (VGI) project itself (with regards to its nature of participation, editing culture/concepts, quality and credibility, etc.). For instance, Mooney and Corcoran

A. Mobasheri (✉)
GIScience Research Group, Heidelberg University, Im Neuenheimer Feld 348, 69120 Heidelberg, Germany
e-mail: a.mobasheri@uni-heidelberg.de

A. Mobasheri (ed.), *Open Source Geospatial Science for Urban Studies*, Lecture Notes in Intelligent Transportation and Infrastructure,
https://doi.org/10.1007/978-3-030-58232-6_1

[1] study the concept of tagging and annotation in OSM by analyzing 25,000 heavily-edited objects belonging to four countries in order to understand the way contributors tag objects in OSM, and further define the characteristics that these objects have [2]. Other studies discuss the motivation of participation [3] and the activity patterns [4] adapted by the volunteers of OSM. Such research studies have tremendously helped the projects to evolve and mature, to an extent that they can replace official proprietary datasets collected by mapping agencies, to an acceptable level.

Numerous projects have employed extractions of data from OSM depending on their needs, mostly because OSM provides up-to-date geographic information that is hard/expensive to access/collect from proprietary datasets. One of the important application domains that open geospatial data has proven to be useful for is urban studies. Such applications vary from car routing and navigation [5, 6], E-bike urban navigation [7] and wheelchair routing and navigation [8, 9], to disaster management [10–12], as well as other urban analytical studies [13–15]. The studies show how important crowdsourced geographic information and in general open geospatial datasets are [16], even though there are concerns with regards to its quality and credibility [17, 18]. Apart from OSM, there are other numerous urban research studies that benefit from using other sources of crowdsourced data such as Twitter [19–21], Flicker [22, 23], and Foursquare [24–26] which most of the times include geographical coordinates.

In addition to crowdsourced geo-data sources that are openly available for usage, several governments have ensured resources for building and setting up open data repositories where most of the published open licensed data contain geospatial components. A promising example of such open data repositories is the open data portal of Austria which at the time of this writing, includes more than 26,000 datasets for nearly 500 application domains from 1184 organizations. The cumulated datasets from all governments around the globe stands as a treasure mine for research and industrial projects. There are several examples of governmental and European funded projects where datasets from open data platforms have been exploited for everyday life applications. As an example, CAP4Access European project (running from 2014 to 17) aimed at employing open geospatial data (including crowdsourced data) for improving urban accessibility. Within this project, OSM data was explored regarding its suitability for Wheelchair routing and navigation [27], and further enriched [28, 29] in order to become qualified for wheelchair routing and navigation. Not just OSM data, but Kerb data of Vienna from Austria's open data portal was used in this research project for means of integrating it with OSM data for improving the routing service. Other similar research projects include Ecocity Tools [30], Road-Kill [31], and COBWEB [32], to name a few. This demonstrates the added value and benefits that openness of geospatial data has brought into such European-wide studies/projects.

The objective of this book and this chapter is to introduce the importance and values of open source geospatial science particularly in urban studies and the following sections provide information on the recent contributions in this domain.

2 Trends in Open Source Geospatial Science for Transportation Studies

The next three chapters of this book focus on the employment of open geospatial datasets and solutions for addressing a research challenge in urban transportation. In this section a brief introduction of these chapters are provided.

Bicycles have an important and positive effect on people who live in urban areas since they provide not only relief of traffic congestion but also enhance the health of citizens. The determination of suitable locations of bicycle sharing system stations and bicycle lanes has drawn attention because of their contributions on bicycles being part of everyday life. In Chapter "Bicycle Station and Lane Location Selection Using Open Source GIS Technology", a workflow that integrates GIS and MCDM methods for the determination of locations of bicycle sharing system stations and bicycle lanes is introduced. For this purpose, MCDM methods are used to identify which criterion is being more effective than others since each criterion affects the location selection process with a different amount. To provide a more useful and reproducible solution, the site selection model is prepared in a widely-used open source GIS software: QGIS. First, three different suitability indexes are obtained by using the weights from MCDM methods. Afterwards, the average analysis is applied to these suitability indexes so as to find the final suitability index and increase the reliability of the result. Furthermore, three different scenarios that take into consideration whether the study area currently has a bicycle sharing system station and/or bike lane are implemented. Various alternative locations for bicycle sharing system stations and bike lanes are proposed in order to support urban planning studies.

As another study of employing open source GIS for transportation purposes, Chapter "Spatial Query Performance Analyses on a Big Taxi Trip Origin–Destination Dataset" develops an open-source library to analyze the query performance of two renowned DBMS: PostgreSQL/PostGIS and MongoDB, a relational and NoSQL DBMS respectively. An open-source Python library is developed to facilitate systematic performance analyses between these DBMSs. The experiments are carried out on New York City's openly available taxi trip origin–destination dataset. The performance of two spatial queries (k-nearest neighbor and point-in-polygon) are investigated in terms of run-time and spatial accuracy. The results indicate the superiority of MongoDB. It outperformed PostgreSQL in terms of run-time in both of the investigated queries. In addition, it is more accurate in terms of detecting k-nearest neighbors. The developed open-source library is utilized to investigate journey time variations between two airports of New York City, which demonstrates its effectiveness in terms of teaching DBMS or GIS modules.

Furthermore, one of the main concerns in employing open geospatial data for urban transportation is its suitability and credibility of use. Volunteered Geographic Information in general and OpenStreetMap dataset in particular is subject to errors, and the it is only reliable to use such datasets for transportation studies only and only if the geographical positions of data features (e.g. roads) are accurate. The current OpenStreetMap practice is that the last node location edit of a map-feature is presented in the map, possibly inferring that it is the most accurate representation. Accordingly, OpenStreetMap presents the recent version of the accumulated edited and mapped data, disregarding the earlier versions. Chapter "Investigating the Use of Historical Node Location Data as a Source to Improve OpenStreetMap Position Quality" investigates the positional quality of historical node versions, evaluating the current OpenStreetMap practice alongside alternative spatiotemporal data models for improved node location calculation. With the proper incremental use of the historical node versions, the more accurate location for close to 60% of all the analyzed nodes was automatically selected between the last version location and the location calculated by the best model. This statistic increases to 76% for nodes that show larger discrepancies when compared to the reference locations. These preliminary results challenge the current OpenStreetMap practice by suggesting an alternative data model that use historical versions. The outcome validates and extends the existing patchwork approach associated with the crowdsourcing of user-generated geographic data and information that are the basis of OpenStreetMap.

3 Trends in Open Source Geospatial Science for Smart Cities

The next four chapters of this book focus on the employment of open geospatial solutions for addressing a research challenge in smart city analytics. In this section a brief introduction of these chapters are provided.

In recent years, there is a widespread growth of smart cities. These cities aim to increase the quality of life for its citizens, making living in an urban space more attractive, livelier, and greener. In order to accomplish these goals, physical sensors are deployed throughout the city to oversee numerous features such as environmental parameters, traffic, and the resource consumption. However, this concept lacks the human dimension within an urban context, not reflecting how humans perceive their environment and the city's services. In this context there is a need to consider sentiment analysis within a smart city as a key element toward coherent decision making, since it is important not only to assess what people are doing, but also, why they are behaving in a certain way. In this sense, the work presented in Chapter "Open Geospatial Data Contribution Towards Sentiment Analysis Within the Human Dimension of Smart Cities" aims to assemble tools and methods that can collect, analyze and share information, based on user-generated spatial content and Open Source geospatial science. The emotional states of citizens were sensed through social media data sources (Twitter), by extracting features (location, user profile information and tweet content by using the Twitter Streaming API) and applying machine learning techniques, such as natural language processing (Tweepy 3.0, Python library), text analysis and computational linguistics (Textblob, Python library). With this approach we are capable to map abstract concepts like sentiment while linking both quantitative and qualitative analysis in human geography. This work would lead to understand and evaluate the "immaterial" and emotional dimension of the city and its spatial expression, where location-based social networks, can be established as pivotal geospatial data sources revealing the pulse of the city.

Furthermore, in the past years, many cities, states, and countries have provided or are currently launching the provision of free and open geodata through public data portals, web-services, and APIs that are suitable for urban and smart cities applications. Besides ready to use 3D city models, many free and open LiDAR data sets are available. Several countries provide national LiDAR datasets of varying coverage and quality as free and open data. In Chapter "Generating 3D City Models from Open LiDAR Point Clouds: Advancing Towards Smart City Applications", a novel pipeline to generate standardized CityGML conform Level of Detail (LoD)-2 city models for city-wide applications by using LiDAR generated point clouds and footprint polygons available from free and open data portals is presented. The proposed method identifies the buildings and rooftop surfaces inside each footprint and classifies them into one of the five rooftop categories. When multiple buildings are present inside a footprint, it is divided into their corresponding zones using a novel corner-based outline generalization algorithm, addressing the need for more precise footprints and models in geometric and semantic terms. Finally, CityGML 2.0 models are created according to the selected category. This pipeline was tested and evaluated on a point cloud dataset which represent the urban area of the Spanish city of Logroño. The results show the effectiveness of the methodology in determining the rooftop category and the accuracy of the generated CityGML models.

Location cover models aim to site one or more facilities in order to provide service to demand in an efficient way. These models are based on economic and location theory. They figure prominently in public and private sector planning, management, and decision-making. Facility location modelling is critically important for both public- and private-sector application, planning, and decision-making contexts for the vision of a smarter city. In the public sector, service facility location selection must account for not only financial costs but also social benefits and urban accessibility. For example, local governments need to determine the locations of public libraries and schools in order to maximise accessibility to/for the public. In the private sector, retail providers must consider store siting that accounts for a range of costs, including transportation to/from distribution centres, delivery and access to serve the target customer base and revenue potential. Because of broad application and extension, these models have been implemented in several geographic information system-based software packages, both proprietary and open source. Among them, open-source implementations are appealing to many because of transparency and cost-effectiveness. As usage among such approaches increases, important questions of optimality and efficiency arise, and heretofore have not been investigated. In general, there is a lack of systematic review of open-source software that supports location coverage modelling. To examine the implications of open-source approaches, a comparative assessment of the functionality provided by open-source packages enabling access to location cover models is provided in Chapter "Open-Source Approaches for Location Coverage Modelling". This study also suggests directions for future implementation improvement.

Finally, Chapter "New Age of Crisis Management with Social Media" of this book tackles a very interesting and challenging application of open geospatial crowd-sourced datasets in urban studies: Disaster/Crisis management. In recent years, Social Media (SM) Volunteered Geographic Information (VGI) is gradually being used for representing the real-time situation during emergency. This chapter presents the SM-VGI review as a new age contribution to emergency management. The study analyses a series of emergencies during the so-called coup attempt within the boundary of Istanbul on the 15th of July 2016 in terms of spatial clusters in time and textual frequencies within 24 h. The aim of the study is to gain an understanding of the usefulness of geo-referenced Social Media Data (SMD) in monitoring emergencies. Inferences exhibit that SM-VGI can rapidly provide the information in the spatiotemporal context with the proper validations, in this way it has advantages to use during emergencies. It is argued that even though geo-referenced data embody the small percent of the total volume of the SMD, it would specify reliable spatial clusters for the events, monitoring with optimized-hot-spot analysis and with the word frequencies of its attributes.

4 Conclusion and Remarks

The majority of the world's population lives in cities, increasing the burden on energy, transport, water supply, construction and public places. There is a growing need for smart urban solutions that benefit from geospatial technologies and are both effective and sustainable. Sustainability in solutions and technologies can be addressed by employing open source datasets and implementing open standards since such studies are reproducible and extensible. Hence, the need for research on open source geospatial science for urban studies is inevitable. The studies presented in this book are examples of recent trends in using open geospatial data as well as open standards and technologies for two main themes of urban studies: transportation and smart city analytics.

Chapters "Bicycle Station and Lane Location Selection Using Open Source GIS Technology" to "Open-Source Approaches for Location Coverage Modelling" present detailed analyses of a proposed methodology to address an urban challenge while stressing the relevancy and importance of open source GIS concepts. While each chapter has its own research objectives, an important outcome of all is the proven fact that open source geospatial technologies can properly replace old-fashion solutions and even in some cases can better address the relevant urban challenges. It is important to note that a main benefit of using open source GIS in these examples is the chance that is given to others to reproduce and implement the same in their city/country. This is key for development of under-developed cities that have less budget compared to first-class cities and country capitals.

Last but not the least, in Chapter "New Age of Crisis Management with Social Media", a comprehensive review of VGI concept and technology in crisis management is presented. The study shows the unique and unquestionable value that open geospatial data and concepts bring in solving disaster urban challenges, exemplified with various studies. Proper management of crisis requires access to up-to-date detailed geographic information, and VGI has proven to be capable of providing this information to the decision-makers up to an acceptable degree, while remaining challenges do exist to be addressed in future studies.

Geo-spatial analysis is a vital part of many scientific studies, particularly in the fields of urban development. Open Source GIS is a useful tool of increasing utility for spatial visualization and analysis, making it a viable alternative to other expensive software packages. Open-source software packages as well as open data-sources benefit from the contributions of experts and users worldwide. The open-source nature of such products provides an opportunity for urban scientists and city-planners to learn and incorporate GIS technology in their studies. The examples covered in this book only scratch the surface of such technologies and their capabilities. Hopefully, the studies provided here can serve as a starting point for your exploration into the functionality and usability of open source geospatial science in urban studies.

References

1. Mooney, P., Corcoran, P.: The annotation process in OpenStreetMap. Trans. GIS **16**(4), 561–579 (2012a)
2. Mooney, P., Corcoran, P.: Characteristics of heavily edited objects in OpenStreetMap. Future Internet **4**(1), 285–305 (2012b)
3. Budhathoki, N.R., Haythornthwaite, C.: Motivation for open collaboration: crowd and community models and the case of OpenStreetMap. Am. Behav. Sci. **57**(5), 548–575 (2013)
4. Neis, P., Zipf, A.: Analyzing the contributor activity of a volunteered geographic information project—the case of OpenStreetMap. ISPRS Int. J. Geo-Inf. **1**(2), 146–165 (2012)
5. Suger, B., Burgard, W.: Global outer-urban navigation with openstreetmap. In: 2017 IEEE International Conference on Robotics and Automation (ICRA), May 2017, pp. 1417–1422. IEEE
6. Graser, A., Straub, M., Dragaschnig, M.: Is OSM good enough for vehicle routing? A study comparing street networks in Vienna. In: Progress in Location-Based Services 2014, pp. 3–17. Springer, Cham (2015)
7. Wei, C.C., Lin, J.S., Chang, C.C., Huang, Y.F., Lin, C.B.: The development of E-bike navigation technology based on an OpenStreetMap. Smart Sci. **6**(1), 29–35 (2018)
8. Menkens, C., Sussmann, J., Al-Ali, M., Breitsameter, E., Frtunik, J., Nendel, T., Schneiderbauer, T.: EasyWheel—a mobile social navigation and support system for wheelchair users. In: 2011 Eighth International Conference on Information Technology: New Generations, Apr 2011, pp. 859–866. IEEE
9. Zipf, A., Mobasheri, A., Rousell, A., Hahmann, S.: Crowdsourcing for individual needs—the case of routing and navigation for mobility-impaired persons. In: European Handbook of Crowdsourced Geographic Information, p. 325 (2016)
10. Soden, R., Palen, L.: From crowdsourced mapping to community mapping: the post-earthquake work of OpenStreetMap Haiti. In: COOP 2014-Proceedings of the 11th International Conference on the Design of Cooperative Systems, Nice (France), 27–30 May 2014, pp. 311–326. Springer, Cham (2014)
11. Dransch, D., Poser, K., Fohringer, J., Lucas, C.: Volunteered geographic information for disaster management. In: Crisis Management: Concepts, Methodologies, Tools, and Applications, pp. 477–496. IGI Global (2014)

12. Mirbabaie, M., Stieglitz, S., Volkeri, S.: Volunteered geographic information and its implications for disaster management. In: 2016 49th Hawaii International Conference on System Sciences (HICSS), Jan 2016, pp. 207–216. IEEE
13. Liu, X., Long, Y.: Automated identification and characterization of parcels with OpenStreetMap and points of interest. Environ. Plan. B Plan. Des. **43**(2), 341–360 (2016)
14. Sun, Y., Du, Y.: Big data and sustainable cities: applications of new and emerging forms of geospatial data in urban studies (2017)
15. Long, Y., Liu, L.: Transformations of urban studies and planning in the big/open data era: a review. Int. J. Image Data Fusion **7**(4), 295–308 (2016)
16. Minghini, M., Mobasheri, A., Rautenbach, V., et al.: Geospatial openness: from software to standards & data. Open Geospat. Data Softw. Stand. **5**, 1 (2020). https://doi.org/10.1186/s40965-020-0074-y
17. Goodchild, M.F., Li, L.: Assuring the quality of volunteered geographic information. Spat. Stat. **1**, 110–120 (2012)
18. Mocnik, F.B., Mobasheri, A., Griesbaum, L., Eckle, M., Jacobs, C., Klonner, C.: A grounding-based ontology of data quality measures. J. Spat. Inf. Sci. **2018**(16), 1–25 (2018)
19. Frias-Martinez, V., Soto, V., Hohwald, H., Frias-Martinez, E.: Characterizing urban landscapes using geolocated tweets. In: 2012 International Conference on Privacy, Security, Risk and Trust and 2012 International Conference on Social Computing, Sept 2012, pp. 239–248. IEEE
20. Jurdak, R., Zhao, K., Liu, J., AbouJaoude, M., Cameron, M., Newth, D.: Understanding human mobility from Twitter. PLoS ONE **10**(7), e0131469 (2015)
21. Roberts, H.V.: Using Twitter data in urban green space research. Appl. Geogr. **81**, 13–20 (2017)
22. Hollenstein, L., Purves, R.: Exploring place through user-generated content: using Flickr tags to describe city cores. J. Spat. Inf. Sci. **2010**(1), 21–48 (2010)
23. Sun, Y., Fan, H., Bakillah, M., Zipf, A.: Road-based travel recommendation using geo-tagged images. Comput. Environ. Urban Syst. **53**, 110–122 (2015)
24. Lindqvist, J., Cranshaw, J., Wiese, J., Hong, J., Zimmerman, J.: I'm the mayor of my house: examining why people use foursquare—a social-driven location sharing application. In: Proceedings of the SIGCHI Conference on Human Factors in Computing Systems, May 2011, pp. 2409–2418. ACM
25. Noulas, A., Mascolo, C., Frias-Martinez, E.: Exploiting foursquare and cellular data to infer user activity in urban environments. In: 2013 IEEE 14th International Conference on Mobile Data Management, June 2013, vol. 1, pp. 167–176. IEEE
26. Humphreys, L., Liao, T.: Foursquare and the parochialization of public space. First Monday **18**(11) (2013)
27. Mobasheri, A., Sun, Y., Loos, L., Ali, A.: Are crowdsourced datasets suitable for specialized routing services? Case study of OpenStreetMap for routing of people with limited mobility. Sustainability **9**(6), 997 (2017)
28. Bakillah, M., Mobasheri, A., Rousell, A., Hahmann, S., Jokar, J., Liang, S.H.: Toward a collective tagging Android application for gathering accessibility-related geospatial data in European cities. Parameters **10**, 21 (2014)
29. Mobasheri, A., Zipf, A., Francis, L.: OpenStreetMap data quality enrichment through awareness raising and collective action tools—experiences from a European project. Geo-Spat. Inf. Sci. **21**(3), 234–246 (2018)
30. Kasprzyk, J.P., Nys, G.A., Billen, R.: Integration of multiple sensor data into a 3D GIS for cities monitoring (2019)
31. Heigl, F., Stretz, C., Steiner, W., Suppan, F., Bauer, T., Laaha, G., Zaller, J.: Comparing road-kill datasets from hunters and citizen scientists in a landscape context. Remote Sens. **8**(10), 832 (2016)
32. de Reyna, M.A., Simoes, J.: Empowering citizen science through free and open source GIS. Open Geospat. Data Softw. Stand. **1**(1), 7 (2016)

Bicycle Station and Lane Location Selection Using Open Source GIS Technology

Dogus Guler and Tahsin Yomralioglu

Abstract To create more sustainable and livable cities, researchers work on different topics. In this context, bicycles have an important positive effect on people living in urban areas since they provide not only relief of traffic congestion but also enhance citizens' health. The finding suitable locations of bicycle sharing system stations and bicycle lanes are attracted attention because they have a huge contribution to providing bicycles are part of everyday life. The aim of this study is to propose a workflow that combines GIS and MCDM methods to determine locations of bicycle sharing system stations and bicycle lanes together. MCDM methods are used to identify which criterion more effective than others since different factors affect the location selection process. Weights of criteria are obtained using AHP, FAHP, and BWM while TOPSIS is applied to rank alternative locations. To provide a more useful and sharable solution, site selection model is prepared in QGIS which is a widely used open source GIS software. First, three different suitability index are obtained using weights that came from MCDM methods. After, average analysis is applied to these suitability indexes so as to increase the reliability of the result. Furthermore, three different scenario applications that take into consideration whether study area has bicycle sharing system station and bike lane currently are implemented in this study. Various alternative locations for bicycle sharing system station and bike lane are proposed in order to support urban planning studies.

Keywords Bicycle sharing system station · Bicycle lane · Best worst method (BWM) · Multi-criteria decision making (MCDM) · Geographic information systems (GIS) · Fuzzy logic

D. Guler (✉) · T. Yomralioglu
Department of Geomatics Engineering, Istanbul Technical University, Istanbul, Turkey
e-mail: gulerdo@itu.edu.tr

T. Yomralioglu
e-mail: tahsin@itu.edu.tr

A. Mobasheri (ed.), *Open Source Geospatial Science for Urban Studies*, Lecture Notes in Intelligent Transportation and Infrastructure,
https://doi.org/10.1007/978-3-030-58232-6_2

1 Introduction

Public transportation systems have high importance for achieving urban sustainability since these systems can reduce traffic congestion, provide fertile energy consumption, and decrease carbon footprints. This is why traffic flow plays an important role in supplying efficient urban economic growth [1, 2]. Motorized vehicles that burn fossil fuels are used as the primary urban transportation mode in order to meet public demand because of fast population growth especially in developing countries [3, 4]. Nevertheless, this causes negative impacts on the environment as these vehicles increase harmful greenhouse gas emissions and exhaust natural resources [5]. To prevent these kinds of negative impacts and secure sustainable urban transportation, urban planners and transportation policymakers try to find a solution by promoting green and efficient public transportation modes that can replace motorized vehicles in urban areas [6, 7]. The increase in demand for green transportation not only contributes to the air quality of cities but also provides active mobility. In this connection, cycling that positively affects the environment and highly contributes to the increase in the quality of persons' health is accepted as one of the most efficient public transportation modes in cities. With the significant planning activities in cities, Bicycle Sharing System (BSS) is an un-ignorable option in order to raise convenience and encourage the use of bicycles [8]. These systems have been operated in over 855 cities worldwide since their first generation showed up in Amsterdam in 1965 [9]. Today, multiple major cities that aim to enable sustainable urban development start new BSS programs around the world. Therefore, new study topics and research are arisen related BSS because of the fast technological developments [10].

The planning of BSS is a complex problem that involves a lot of factors. Primarily, the determination of optimum BSS station numbers and locations are needed in order to enable efficient BSS [11]. Station density that provides ease usage is necessary to increase the number of users [12]. Also, BSS stations located in close proximity to the public transportation stations are of importance with regard to accessibility [13]. In relation to this, Bicycle Lane (BL) is an important element in order to allow effective BSS. For this reason, municipalities try to implement the urban plans that contain new BL in order to increase roadway safety, growing bicycle use, and enhance public health. Nonetheless, cyclists commonly ride on insecure roads that do not include any BL. This brings that cyclists face with a superior risk of crashing [14]. Relatedly, studies show that the safety of BL is a vital concern for cyclists and one of the reasons for the low-density cycling usage is that cycling is not safe enough [15, 16]. Suitable locations of BSS station and BL should be detected by taking the safety factor into consideration. An integrated approach that identifies suitable locations of BSS and BL can be more effective in the context of smart urban planning.

The recent studies related to cycling cover user behavior [13, 17–23], spatial distribution [8, 12, 24–28], spatial equity [1, 12, 29, 30], and safety [14, 31–33]. Many researchers used optimization models and mathematical programming to determine distributions of BSS and BL. For example, the authors proposed a model that contains risk, comfort, service coverage, and impact objectives in order to identify a new bikeway as a case study in Taipei City in [34]. Researchers used a grey 0–1 programming problem and they considered different constraints in their proposed model. Also, they conducted a scenario analysis in terms of landscape and safety. In another study, the authors developed a model that contains multi-objective to determine locations of bikeways and BSS stations in [8]. Their results indicated that a high budget for bikeways enhances the safety and comfort of cyclists. Additionally, this study is one of the few studies that aim to determine optimal locations BL and BSS at the same time. Furthermore, there is a number of studies on cycling that benefit from Geographic Information System (GIS) which are detailed as follows. The authors conducted research that aims to evaluate the accessibility performance of the bicycle network using GIS in Baltimore, Maryland in [35]. They indicated that study results can contribute to land use planning in terms of spatial equity. Researchers determined new bicycle parking locations using a GIS-based approach that considers multiple criteria in [36]. There are GIS-based studies that utilize the grid-cell model [37], demand-based multiple criteria [38], location-allocation model [39] in order to find optimal locations of BSS station and BL. For example, the authors applied a methodology that integrates scaling approach and GIS to find suitable bicycle paths in which the consistency of decisions, however, could not be checked in [40]. In another study, researchers aimed to obtain alternative locations of BSS stations by using multiple criteria and GIS in [41]. Here, the authors utilized kernel density spatial analysis rather than fuzzy logic to normalize criteria.

The researchers frequently benefit from open data and open source geospatial technologies to analyze and improve the use of bicycle, for example assessing of air pollution exposure [42, 43], examining environmental characteristics [44], comparing crowdsourced and conventional cycling datasets [45], examining use of urban reserves [46], exploring spatial behavior of cyclists [47], helping transport decisions [48].

This study outlines an approach that integrates Multi-Criteria Decision Making (MCDM) and fuzzy GIS in order to address the problem of where to build BSS station and BL. This research can contribute to the existing literature;

- To find suitable locations of BSS station and BL simultaneously by using GIS which promotes effective land use planning readily.
- To assist decision-makers by creating a reproducible open source GIS model.
- To better express attributes of different criteria that affect location selection of BSS station and BL by preparing the GIS layers using fuzzy logic for the Weighted Linear Combination (WLC).
- To provide a methodology being used independently of the study area containing BSS station and/or BL.

This chapter is organized into five sections. In the second section, the methodology is described. The next section presents a case study. The fourth section discusses the results of the case study analysis. The conclusions are drawn in the final section.

2 Methodology

This study focuses on location selection of BSS station and BL from the open-source GIS point of view which is reusable by different researchers. The workflow includes three different scenarios related to BSS and BL. In order to realize location selection analysis, effective factors that are used in multi-criteria decision making (MCDM) are determined by taking into account the literature review. The spatial database consisting of data layers that belong to criteria are prepared by obtaining from different data sources. To conduct efficient spatial analyses, all layers should have the same coordinate system and pixel value since suitability analysis is realized using raster-based GIS [49]. Criterion layers should also have normalized pixel values depending on their effect on the suitability for the locations of BSS station and BL. This study examined the usage of different fuzzy membership functions so as to obtain suitabilities of criteria accurately. The weight of each criterion for each different scenario is calculated by using different MCDM methods, namely the Analytic Hierarchy Process (AHP), the Fuzzy AHP (FAHP), and Best Worst Method (BWM). The reason for using different methods to calculate the weights of criteria is to improve the stability of decisions. Thus, the shortcomings of methods are able to be eliminated. The use of multiple methods rather than a single method can provide truer criterion weight. Once the suitability calculated by using weights of criteria relative to each method is obtained, the final suitability is calculated as a means of averaging three suitabilities. The selected alternative locations of BSS stations and BL are ranked using the Technique for Order Preference by Similarity to Ideal Solution (TOPSIS) method by taking into consideration normalized criteria values.

2.1 Fuzzy Modeling

In order to deal with the representation of real situations that are very often uncertain, the fuzzy logic theory that allows the imprecision describing of objects is proposed in [50] and researchers further developed the theory in [51, 52]. It is a superset of conventional (Boolean) logic that has conventional evaluations like yes/no or true/false. While fuzzy logic enables to define of intermediate values, the fuzzy set theory allows the object to belong fuzzy sets instead of a crisp set. The fuzzy set theory describes the grade of membership with membership function $\mu_M(x)$ in the universe of discourse X that has a subset M. In the GIS-related studies, the raster map represents the universe of discourse while the element x is a pixel value. The values of $\mu_M(x)$ express that an element fully belongs to the crisp set X for $\mu_M(x) = 1$ and an element does not have any membership for $\mu_M(x) = 0$. The higher pixel values indicate to have more belonging to the crisp set. A membership value can be any number between zero and one. Therefore, the fuzzy set has a rigid boundary.

However, the capacity of the theory allows the transition between full membership and non-membership by providing intermediary membership. This has broad effectiveness for GIS-based operations and spatial analyses including Multi-Criteria Decision Analysis (MCDA) [53]. Membership functions have three general types as S-shaped, linear, and point. The function type that is used in the researches varies depending on the characteristics of spatial phenomena. In this study, the normalization of criteria values is conducted by benefiting from linear and S-function which are used by different studies [51, 54, 55]. Equations (1) and (2) show the increasing and decreasing S-function formulas, respectively while Eq. (3) presents the linear membership function. The parameters a and b represent possible lowest and highest values that describe changes in fuzzy membership for S-function. A linear function has four parameters as a, b, c, and d to identify changes in membership. These functions are performed by using raster-based calculations in GIS. Fuzzy membership functions help to particularly represent attributes of criteria that affect location selection in the normalization process on the contrary of linear scale transformation.

$$\mu_M(x) = \begin{cases} 0 & x < a \\ \sin\left(\frac{x-a}{b-a} \times \frac{\pi}{2}\right) & a \leq x < b \\ 1 & x \geq b \end{cases} \tag{1}$$

$$\mu_M(x) = \begin{cases} 1 & x < a \\ \sin\left(\frac{b-x}{b-a} \times \frac{\pi}{2}\right) & a \leq x < b \\ 0 & x \geq b \end{cases} \tag{2}$$

$$\mu_M(x) = \begin{cases} 0 & x < a \\ \frac{x-a}{b-a} & a \leq x \leq b \\ 1 & b < x < c \\ \frac{d-x}{d-c} & c \leq x \leq d \\ 0 & x > d \end{cases} \tag{3}$$

2.2 *Spatial Decision Support*

Spatial decision support that helps to expand GIS capabilities is frequently utilized in order to improve the performance of different stakeholders such as decision-makers, managers, and citizens when they try to overcome complex spatial problems like site selection. The integration of MDCM and GIS can yield significant results to solve spatial problems because GIS has the capacity to produce maps that contain evaluations of decision-makers while MCDM can tackle with the disagreements of judgments. This research proposes a methodology that includes different MCDM methods to support decision-makers when they face with BSS station and BL location selection problem.

Weighted Linear Combination (WLC) To obtain the location suitability of BSS station and BL, the WLC method is used in this research. This model consists of two components which are criterion weights w_k and value functions $v(a_{ik})$. Equation (4) shows the WLC formula where $V(A_i)$ is the overall suitability value for the ith alternative location and $v(a_{ik})$ is the normalized criterion value which is obtained by using fuzzy membership functions.

$$V(A_i) = \sum_{k=1}^{n} w_k v(a_{ik}) \tag{4}$$

Analytic Hierarchy Process (AHP) AHP that provides complete evaluations of different criteria to achieve a determined goal is one of the most commonly used MCDM methods. This method aims to assist decision-makers for the solution of complex problems by determining each criterion's weight. In addition to this, AHP supplies an evaluation of consistencies of all judgments because it enables to express consistency ratio as a mathematical formula. AHP can be used as a tool that assesses criterion weights of associated criterion map layers to incorporate into GIS. Once criteria weights are obtained, the global suitability values can be calculated using the WLC technique. To do this, the AHP model including objectives, criteria, and alternatives is established. In the second step, the relative importance of each criterion is assigned with pairwise comparisons performed by decision-makers based on a determined scale which consists of numbers between 1 and 9. The scale is utilized to transform judgments into a numerical representation. The pairwise comparison matrix is consistent and reciprocal. In the last step, criteria weights are calculated using the eigenvector principle [56]. The sum of the weights should always be equal to one according to the method.

$$\mathrm{A} = [a_{ij}] = \begin{matrix} C_1 \\ C_2 \\ \vdots \\ C_n \end{matrix} \begin{pmatrix} 1 & a_{12} & \cdots & a_{1n} \\ 1/a_{12} & 1 & \cdots & a_{2n} \\ \vdots & \vdots & \ddots & \vdots \\ 1/a_{1n} & 1/a_{2n} & \cdots & 1 \end{pmatrix} \tag{5}$$

where A = pairwise comparison matrix, w = eigenvector, and $\lambda_{\max}$ = the largest eigenvalue of A. As mentioned before, AHP checks the stabilities of decisions by calculating the Consistency Index (CI) and the Consistency Ratio (CR). According to the theorem, if the CR value is smaller than 0.1, then the comparisons are acceptable; otherwise, the pairwise comparison matrix should be reestablished. Random Index (RI) represents the mean CI value for a certain number of criteria (Table 1). The

Table 1 Average random consistency index (RI) [56]

N	1	2	3	4	5	6	7	8	9	10
RI	0	0	0.52	0.89	1.11	1.25	1.35	1.40	1.45	1.49

formulas are defined as follows:

$$\lambda_{\max} = \sum_{j=1}^{n} a_{ij} \times \frac{w_j}{w_i} \tag{6}$$

$$Aw = \lambda_{\max} w \tag{7}$$

$$\mathrm{CI} = \frac{\lambda_{\max} - n}{n - 1} \tag{8}$$

$$\mathrm{CR} = \frac{\mathrm{CI}}{\mathrm{RI}} \tag{9}$$

Fuzzy Analytic Hierarchy Process (FAHP) FAHP is an extension of the AHP method and it is proposed to capture decision maker's preferences more accurately by taking into consideration fuzzy logic in order to solve complex problems which indicate commonly characteristic of fuzziness. While FAHP allows decision-makers to express their choices more approximate or flexible, it also integrates fuzziness to judgments so as to tackle with sharp preferences. Therefore, a number of studies are conducted using this method with regard to spatial decision problems. This study employs a combination of extent analysis method and the total integral value method described in [57]. In this methodology, W is the normalized weight vector of triangular fuzzy comparison matrix A. In the first step, the fuzzy synthetic extent value is calculated. $\tilde{c}_{xy}$ is denoted as the Triangular Fuzzy Number (TFN) for comparison of x criterion over y criterion in the related pairwise matrix. This can be represented as (l_{xy}, m_{xy}, u_{xy}). According to the theory, the fuzzy synthetic extent value $\tilde{S}_x = (l_x, m_x, u_x)$ for the criterion x is obtained via the following equation:

$$\tilde{S}_x = \sum_{y=1}^{n} c_{xy} \otimes \left[\sum_{k=1}^{n} \sum_{y=1}^{n} \tilde{c}_{ky} \right]^{-1} \quad x = 1, 2, \ldots, n \tag{10}$$

where n is the fuzzy comparison matrix A. After, the fuzzy addition operations are performed as Eqs. (11) and (12):

$$\sum_{y=1}^{n} \tilde{c}_{xy} = \left(\sum_{y=1}^{n} l_{xy}, \sum_{y=1}^{n} m_{xy}, \sum_{y=1}^{n} u_{xy} \right) \quad x = 1, 2, \ldots, n \tag{11}$$

$$\sum_{k=1}^{n} \sum_{y=1}^{n} \tilde{c}_{ky} = \left(\sum_{k=1}^{n} \sum_{y=1}^{n} l_{ky}, \sum_{k=1}^{n} \sum_{y=1}^{n} m_{ky}, \sum_{k=1}^{n} \sum_{y=1}^{n} u_{ky} \right) \tag{12}$$

and then the inverse of the vector in Eq. (12) is computed using Eq. (13):

$$\left[\sum_{k=1}^{n}\sum_{y=1}^{n}\tilde{c}_{ky}\right]^{-1} = \left(\frac{1}{\sum_{k=1}^{n}\sum_{y=1}^{n}u_{ky}}, \frac{1}{\sum_{k=1}^{n}\sum_{y=1}^{n}m_{ky}}, \frac{1}{\sum_{k=1}^{n}\sum_{y=1}^{n}l_{ky}}\right) \quad (13)$$

After this, the synthetic extent values of A are obtained using the total integral value theory.

$$I_T^{\alpha}(\tilde{S}_x) = \frac{1}{2}\alpha(m_x + u_x) + \frac{1}{2}(1-\alpha)(l_x + m_x) = \frac{1}{2}[\alpha u_x + m_x + (1-\alpha)l_x] \quad (14)$$

where α is the index of optimism that indicates a decision maker's optimism level. α is a number between 0 and 1. If this number is closer to 0, the decision is more pessimistic, otherwise; it is optimistic. Finally, the normalized weight vector $W = (w_1, w_2, \ldots, w_n)^T$ is obtained using the following equation:

$$w_x = \frac{I_T^{\alpha}(\tilde{S}_x)}{\sum_{k=1}^{n} I_T^{\alpha}(\tilde{S}_k)} \quad x = 1, 2, \ldots, n \quad (15)$$

where w_x is a non-fuzzy number. The stabilities of decisions are determined using CR as in the AHP theory. Before that, fuzzy numbers are turned to crisp numbers via Eq. (16):

$$M = \frac{l + 4m + u}{6} \quad (16)$$

where M represents transformed crisp numbers from TFN belonging to the comparison matrix A.

Best Worst Method (BWM) BWM is based on the pairwise comparisons as AHP and Analytic Network Process (ANP) methods in order to obtain weights of different criteria that affect the decision and it is one of the most recent MCDM methods [58]. BWM requires less pairwise comparisons than AHP and these comparisons are always consistence because of the theory of the method. This method uses vectors instead of matrices to compose pairwise comparisons. Also, BWM realizes the calculations using integer numbers rather than rational numbers for allowing easier implementation than other methods. These are the main advantages of BWM. In the method, supremacy level of the best criterion over other criteria and supremacy levels of all criteria over the worst criterion is determined by using a number scale between 1 and 9 so as to obtain weights of criteria. The steps of the BWM are as follows [59]:

Step 1. Determine a set of criteria that affect the decision as $\{c_1, c_2, \ldots, c_n\}$.

Step 2. Determine the most desirable criterion as the best and the least as the worst.

Step 3. Determine the comparative degree of the best criterion over all other criteria using the number scale between 1 and 9 in order to compose best-to-others vector $A_B = (a_{B1}, a_{B2}, \ldots, a_{Bn})$ where B is the best criterion and a_{Bj} represents the comparative degree over the criterion j of it. Clearly, $a_{BB} = 1$.

Step 4. Determine the comparative degree of the all criteria over worst criterion using the number scale between 1 and 9 in order to compose others-to-worst vector $A_W = (a_{1W}, a_{2W}, \ldots, a_{nW})^T$ where W is the worst criterion and a_{jW} represents the comparative degree over it of criterion j. Clearly, $a_{WW} = 1$.

Step 5. Obtain the optimal weights $\left(w_1^*, w_2^*, \ldots, w_n^*\right)$.

The finding of the optimal weights can be expressed as the following linear programming problem:

$$\begin{aligned}
&\min \xi^L \\
&\text{s.t.} \\
&\left|w_B - a_{Bj} w_j\right| \le \xi^L, \quad \text{for all } j \\
&\left|w_j - a_{jW} w_W\right| \le \xi^L, \quad \text{for all } j \\
&\sum_j w_j = 1 \\
&w_j \ge 0, \quad \text{for all } j
\end{aligned}$$

The optimal weights $\left(w_1^*, w_2^*, \ldots, w_n^*\right)$ and consistency indicator ξ^{L*} are obtained by solving the model. The value of ξ^{L*} should be close to zero for more consistent results.

Technique for Order Preference by Similarity to the Ideal Solution (TOPSIS)
TOPSIS, originally proposed in [60], is a Multi-Attribute Decision Making (MADM) method that is used to select the best alternative by decision-makers. This simple method based on that best alternative should have a minimum distance to a positive ideal solution and should have a maximum distance to a negative ideal solution. TOPSIS has gained much interest from researchers who conduct studies on different topics. In this study, we use this method so as to obtain a ranking of the selected alternative BSS station and BL. The implementation steps of this technique are as follows:

Step 1. Establish a normalized decision matrix as $r_{ij} = x_{ij} \Big/ \sqrt{\left(\sum X_{ij}^2\right)}$ for $i = 1, \ldots, m$; $j = 1, \ldots n$ where x_{ij} is the original and r_{ij} is the normalized score of the decision matrix.

Step 2. Establish the weighted normalized decision matrix as $v_{ij} = w_i r_{ij}$ where w_i is the weight of criterion j.

Step 3. Determine the positive ideal solution as $A^* = \{v_1^*, \ldots, v_n^*\}$ where $v_i^* = \{\max(v_{ij}) \text{ if } j \in J; \min(v_{ij}) \text{ if } j \in J'\}$ and determine the negative ideal solution as $A' = \{v_1', \ldots, v_n'\}$ where $v' = \{\min(v_{ij}) \text{ if } j \in J; \max(v_{ij}) \text{ if } j \in J'\}$.

Step 4. Calculate the distance from a positive ideal alternative as $S_i^* = \left[\sum \left(v_i^* - v_{ij}\right)^2\right]^{1/2}$ for $i = 1, \ldots, m$ and calculate the distance from the negative ideal alternative as $S_i' = \left[\sum \left(v_j^* - v_{ij}\right)^2\right]^{1/2}$ for $i = 1, \ldots, m$ for each alternative.

Step 5. Calculate the relative proximity to the ideal solution as $C_i^* = S_i' / (S_i^* + S_i')$, clearly, $0 < C_i^* < 1$. Finally, select the alternative having the highest C_i^* value.

2.3 *Open Source Geographic Information Systems Modeling*

In the first place, the general public reaches the new algorithms and codes that are developed by universities and governmental agencies; therefore, open-source software notions can be considered as old as software development itself [61]. The developments of open software increased in the past several years thanks to start of new projects and support of the governmental agencies [62–64]. In addition, it is observed that a significant rise in download rates of open source GIS software [65]. This explains the notable increase in use cases of open source GIS software, for example, water resource analyses [66, 67] and landscape applications [68]. In this context, a number of research papers and books that mention production and usage of open source GIS software tools and libraries are published while plenty of research projects are conducted. Moreover, developed products are published under open source licenses [69]. QGIS project emerged with the aim of providing an effortless user interface to process spatial data in Linux-based systems. After that, the numbers of volunteer developers and users of QGIS rapidly increased and it has one of the largest communities between the open-source GIS software. Additionally, QGIS is distributed under the GNU General Public License and it is an official project of the Open Source Geospatial foundation (OSGeo). Another important open source GIS movement is the SAGA (System for Automated Geo-Scientific Analysis) which is designed to execute vector and raster data analysis and is developed in 2001 at the Department of Geography at the University of Göttingen (Germany). Furthermore, Geospatial Data Abstraction Library (GDAL) that is used to read and write various spatial data formats such as GeoJSON and GeoTIFF is released under the X/MIT style Open Source License by the OSGeo.

Since one of the main aims of this study is to create an automatic fuzzification process for input spatial data regarding effective criteria in the location selection of BSS station and BL, several different models that include spatial analysis features of GDAL, SAGA GIS, and QGIS are established using model tool of QGIS in order to obtain fuzzificated criteria layers. Figures 1 and 2 show the fuzzification tools

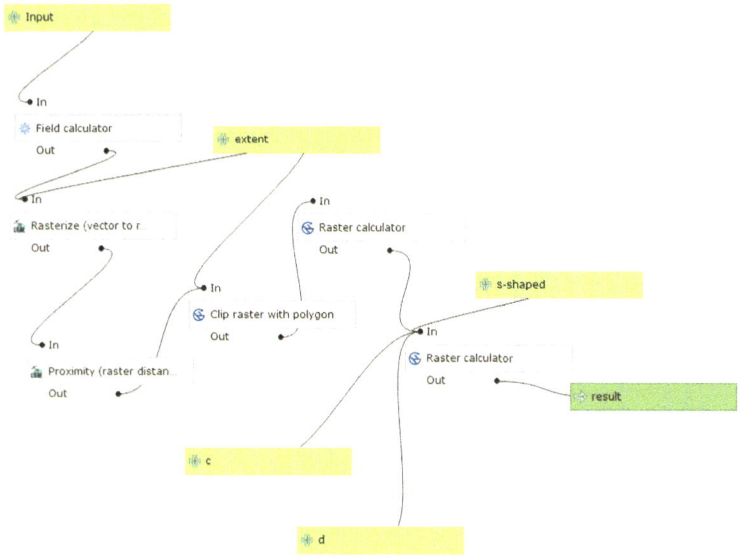

Fig. 1 Fuzzification tool (S-function)

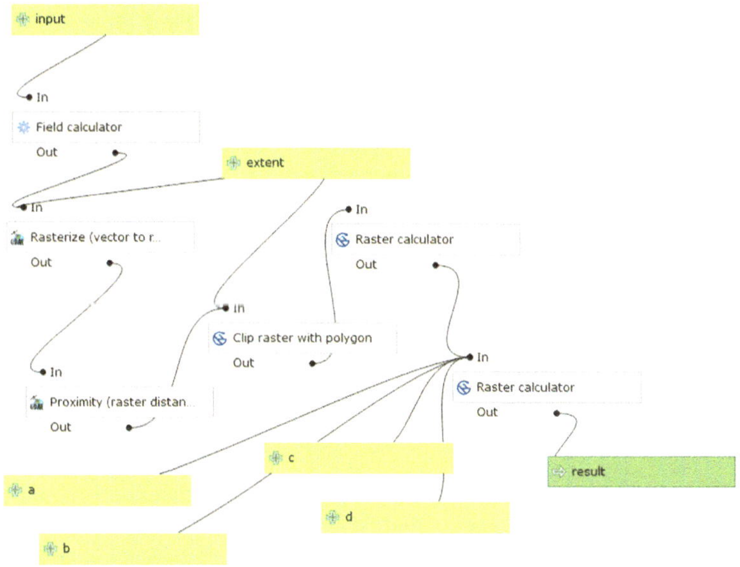

Fig. 2 Fuzzification tool (linear function)

prepared for the S-function and linear function calculations which are discussed in Sect. 2.1.

In models, firstly, the user enters the required data including study extent information, point, line, or polygon data, S-function or linear function information, threshold values. In the first stage, input vector data is converted into proximity raster data and is clipped by using the study extent data. In these processes, the *proximity* analysis tool of GDAL and *clip raster with the polygon* tool of SAGA GIS is executed. After, pixel values of obtained raster data are converted into the integer type. In the last stage, fuzzified pixel values are achieved using input data and conditional formatting features of the model. In this stage, the *raster calculator* tool of SAGA GIS is utilized. The fuzzification models used in this study are published.[1]

3 A Case Study: Istanbul

The proposed methodology is used for six neighborhoods of Istanbul located northwest of Turkey with a shoreline. The city faces with transportation problems because of high rates in population and migration. The population of Istanbul is 15,067,724 according to the Turkish Statistical Institute (TurkStat) data in 2018. Consequently, city administrators aim to actualize new public transportation solutions in order to cope with the problems that hinder to ensure sustainable transportation. According to statistics, the usage percentages of railway, highway, and seaway are 18.07%, 78.15%, and 3.77%, respectively. ISBIKE is the only BSS of Istanbul currently. The system is carried out with 140 stations and 1500 bicycles in total. According to the statement of ISBIKE, the number of bicycles will be 3000 at the end of 2019. These stations are mostly located in shorelines. There are 27 BSS stations and 15 km BL in the study area. These data can be obtained from the ISBIKE website and OpenStreetMap. However, there is no official data about BL shared publicly. Figure 3 illustrates the study area.

3.1 Scenarios

One of the aims of this study is to propose an approach that can be used to find the optimal location of BSS station and BL in different application areas. For this reason, several scenarios are created considering whether application areas include BSS station and BL or not. It is considered that this can contribute to the re-use of the proposed approach by various decision-makers. Here, we can express three scenarios as:

- Scenario 1 (S1): Application area includes neither BSS station nor BL.

[1] https://github.com/gulerdo/spatial-fuzzification.

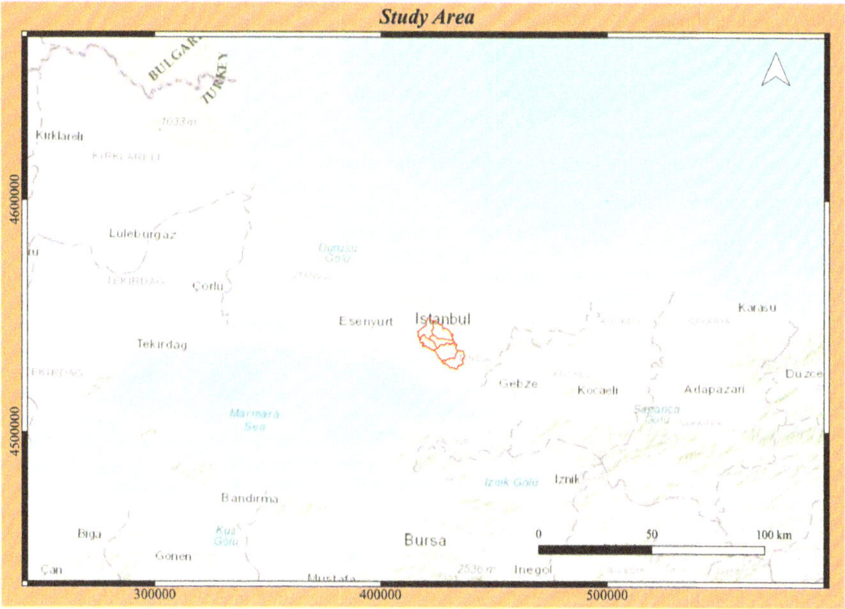

Fig. 3 Map of the study area

- The data of BSS station and BL are not implemented and it is assumed that the case study area does not contain any BSS and BL.

- Scenario 2 (S2): Application area includes BL but it does not include BSS station.
  - The data of BSS station are not implemented and it is assumed that the case study area only contains BL.
- Scenario 3 (S3): Application area includes BSS station and BL together.
  - The data of BSS station and BL are both implemented. So, this scenario represents the reality of the case study area.

3.2 Evaluation of Criteria

Finding the optimal locations of BSS station and BL is a complex decision-making problem because it can be affected by varied factors. Accordingly, related literature is examined and effective criteria are identified in order to fulfill the decision-making process. In this study, nine criteria are used in total. *Proximity to public parks criterion (C1)* enables to use of BSS station and BL for different aims such as transportation and hobby. *Proximity to shopping malls criterion (C2)* can contribute to the usage of BSS station and BL because shopping malls are visited by a number of guests from distances that can be covered by bicycle. *Proximity to bus lines criterion (C3)*

Table 2 Function values

	Minimum Level (0)	Intermediate Interval (0–1)	Optimum Level (1)	Function type
(C_1) Proximity to public parks	>5000	500–5000	<500	S (Decreasing)
(C_2) Proximity to shopping malls	>5000	500–5000	<500	S (Decreasing)
(C_3) Proximity to bus lines	<50	100–150	>750	Linear
(C_4) Proximity to transportation stations	>5000	500–5000	<500	S (Decreasing)
(C_5) Proximity to education facilities	>5000	500–5000	<500	S (Decreasing)
(C_6) Population density	<30,000	30,000–100,000	>100,000	S (Increasing)
(C_7) Slope	>5%	2–5%	<2%	S (Decreasing)
(C_8) Proximity to bike lanes	>5000	100–5000	<100	S (Decreasing)
(C_9) Proximity to bicycle stations	>5000	500–5000	<500	S (Increasing)

is substantial in terms of cyclists' safety and connection to bus transportation. Therefore, BSS station and BL should not be located too close distance to bus lines; on the other hand, they should be located moderate distance from these lines to prevent accidents. *Proximity to transportation stations criterion (C4)* can ensure efficient usage of bicycles for citizens transferring to transportation stations other than bus stations. *Proximity to education facilities criterion (C5)* is a significant factor for great student potential in using bicycles. *Population density criterion (C6)* is a smart indicator that shows the demand for cycling. *Slope criterion (C7)* is related that the roads with high slope are rarely preferred by cyclists. Although usage of *proximity to bike lanes criterion (C8)* varies according to the identified scenario, BL is an important criterion for location selection since it can assist to broad adoption of cycling in terms of connectivity of different BL. *Proximity to bicycle stations criterion (C9)* can affect to finding locations of BSS station and BL since bicycle stations should locate in the moderate distance from existing and new BSS station and BL. Table 2 details the function types and threshold values of all effective criteria. As detailed in Table 2, level (1) represent the values for most suitable locations while level (0) represent the values for least suitable locations; however, interval (0–1) represents the transition values for locations with from least suitable to most suitable. Figure 4 presents the fuzzy values of criteria and Fig. 5 shows the maps of criteria.

3.3 *Fuzzification of Layers*

After the identification of criteria, spatial layers that represent related criteria are prepared by using different resources such as OpenStreetMap, Istanbul Metropolitan Municipality, Turkish Statistical Institute, ISBIKE, and EUDEM.

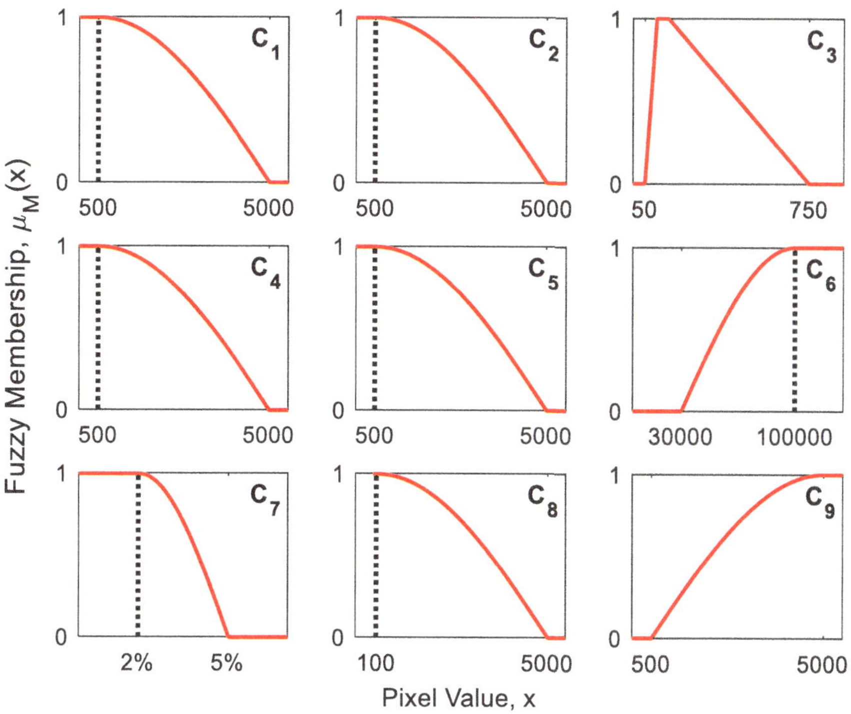

Fig. 4 Fuzzy membership values

In order to make the WLC process flawlessly, all criteria layers should be normalized into the same scale. In the next step, criterion layers are converted to normalized raster data that represent suitability for location selection analysis by using fuzzification tools discussed in Sect. 2.3. Figure 6 presents the fuzzified maps of criteria. The pixel numbers of each map are between 0 and 1.

3.4 Criteria Weighting

The determination of criteria weights is one of the most important steps for MCDM studies due to the fact that it affects the results directly. Criteria weights for different scenarios are determined by the authors by evaluating the pairwise comparisons in [38–41, 70]. Table 3 lists all criteria weights that are calculated by implementing the process steps of the methods mentioned in Sect. 2.2. Also, consistencies of all pairwise comparisons that express the judgments are calculated. As can be seen in Table 3, all consistency values are small than 0.1; hence obtained criteria weights are

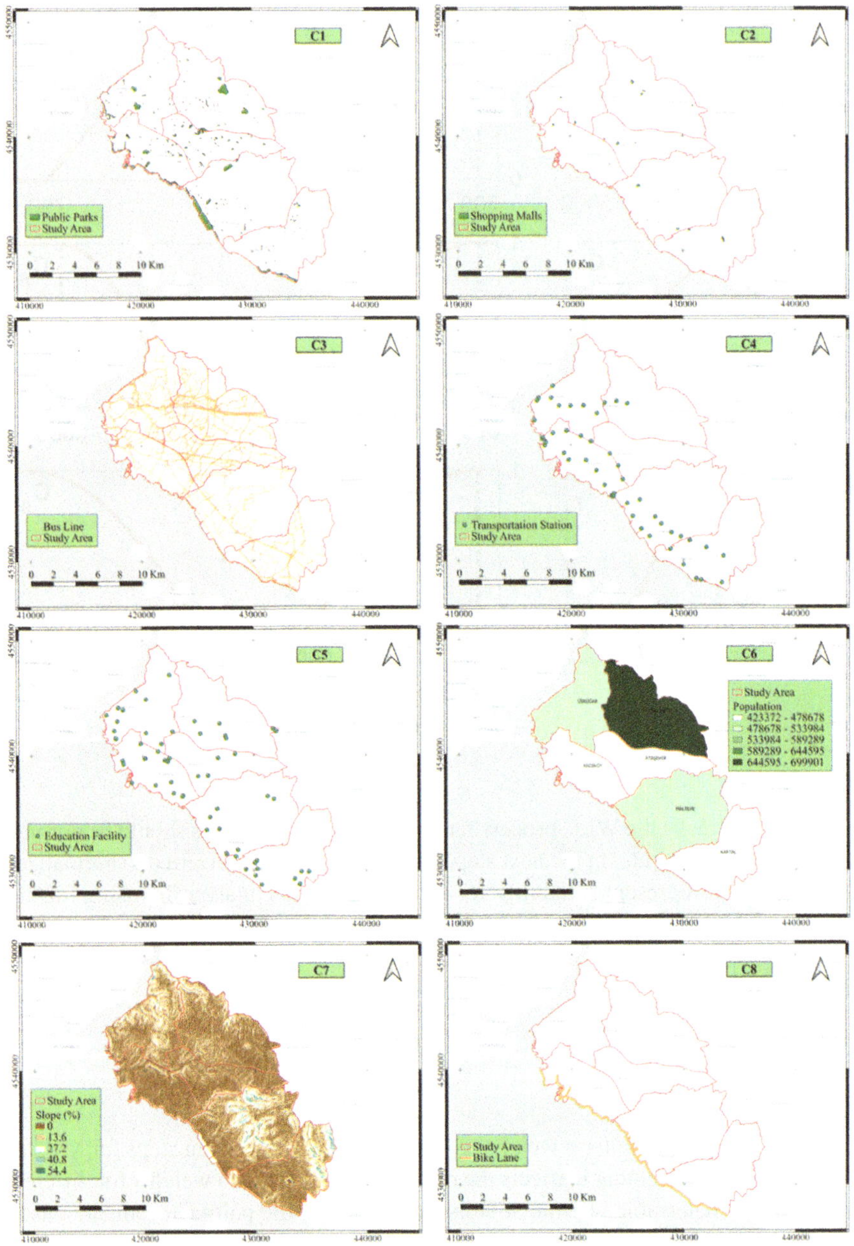

Fig. 5 Maps of the criteria

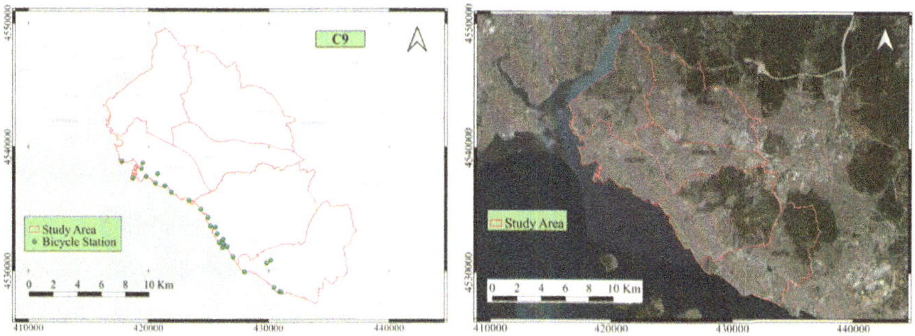

Fig. 5 (continued)

acceptable. When Table 3 is examined, it can be seen that the most important criterion is C4 while the least important criterion is C2. This supports enabling cycling as one of the transportation ways.

4 Results and Discussion

After the criterion layers are prepared, results are obtained by utilizing the WLC process. In this sense, the *raster calculator* analysis tool is used in the proposed approach. Each normalized criterion layer is associated with its weight. Since each criterion has three different weights, the average values of all obtained data are calculated. So, the final suitability index is obtained. This whole process is utilized for each aforementioned scenario. Following, all scenario result data are classified into five equal interval suitability classes as strong unsuitable (0–0.2), slightly unsuitable (0.2–0.4), slightly suitable (0.4–0.6), suitable (0.6–0.8), and strong suitable (0.8–1). Alternative BSS stations are selected by taking into consideration the pixel values of suitability result data for each scenario. It is considered that each station should have a minimum of 250 m distance from other stations when selecting the alternatives. Also, it is ensured that current stations have at least 100 m distance from alternative stations by using Delaunay triangulation. Pixel values of obtained suitability data are associated with the road network of the study area by using add *raster values to features* tool of SAGA GIS in order to identify alternative BL. The data are classified according to their suitability pixel values. Subsequently, alternative BL is created by taking into consideration suitable and strong suitable classes. It is considered that alternative BL can connect with current BL. Figures 7, 8 and 9 show the alternative BSS stations and BL for three different scenarios. Three scenarios have 39, 34, and 27 alternative BSS stations, respectively. This may arise from that S1 needs new facilities more than other scenarios because it has not any BSS station and BL. Furthermore, scenarios have 6, 5, and 5 alternative BL, respectively. As illustrated

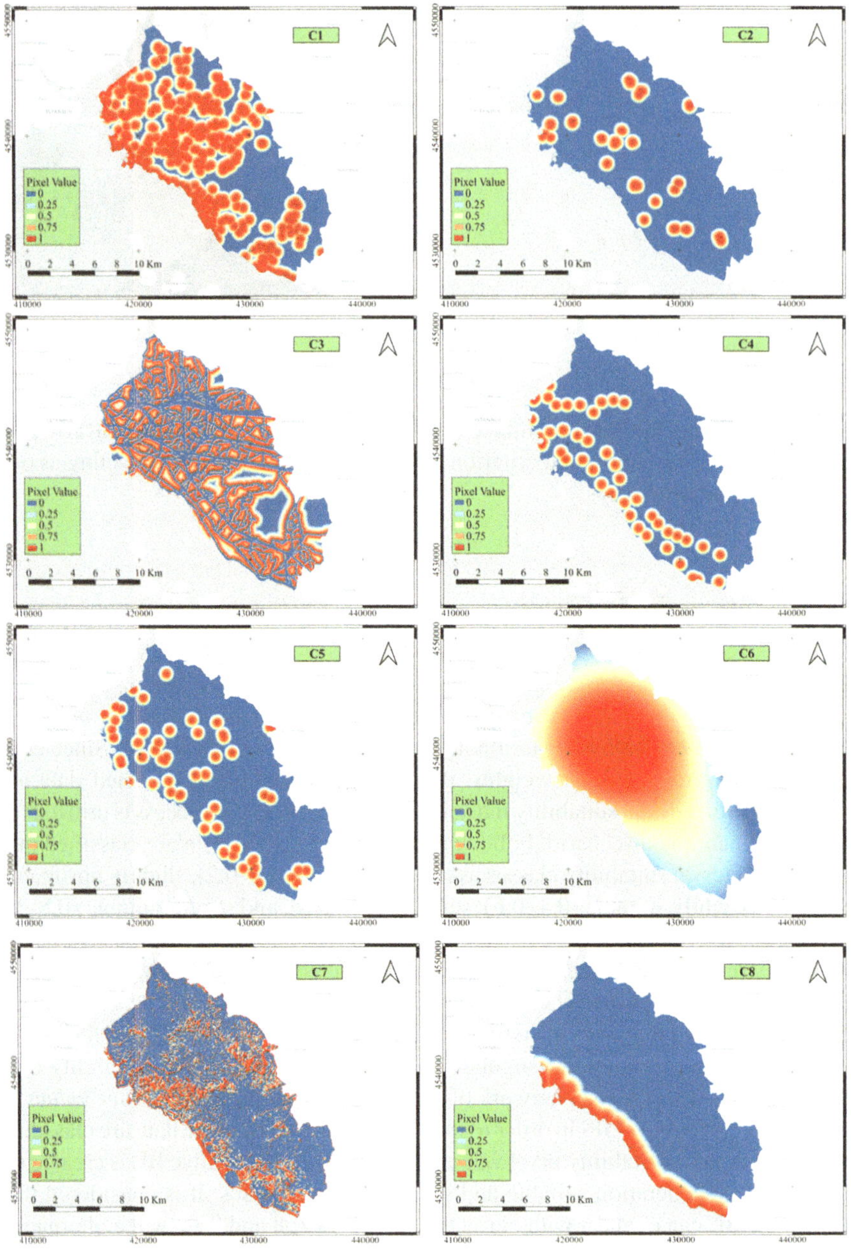

Fig. 6 Maps of fuzzificated criteria

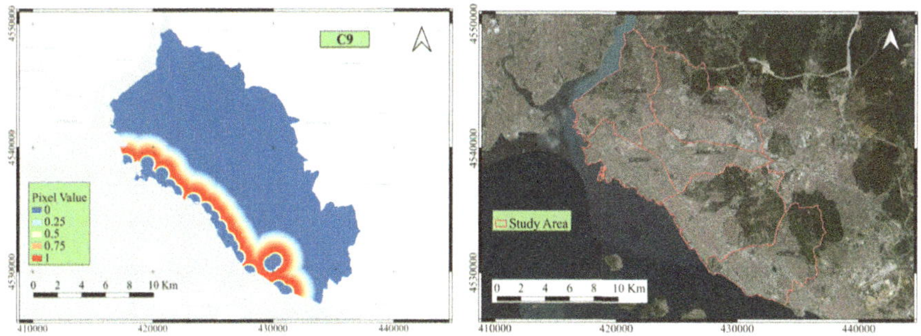

Fig. 6 (continued)

Table 3 Criteria weights

	S1			S2			S3		
	AHP	FAHP	BWM	AHP	FAHP	BWM	AHP	FAHP	BWM
(C1)	0.1887	0.1559	0.1589	0.1766	0.1473	0.1419	0.1611	0.1423	0.1283
(C2)	0.0300	0.0293	0.0371	0.0257	0.0258	0.0331	0.0223	0.0236	0.0299
(C3)	0.1310	0.1505	0.1589	0.1300	0.1345	0.1419	0.1226	0.1250	0.1283
(C4)	0.3303	0.2977	0.3865	0.2913	0.2652	0.3454	0.2719	0.2384	0.3122
(C5)	0.0761	0.1019	0.0953	0.0636	0.0819	0.0852	0.0555	0.0713	0.0770
(C6)	0.0355	0.0614	0.0681	0.0296	0.0497	0.0608	0.0260	0.0432	0.0550
(C7)	0.2083	0.2034	0.0953	0.1935	0.1839	0.0852	0.1748	0.1652	0.0770
(C8)				0.0897	0.1119	0.1065	0.0806	0.0958	0.0962
(C9)							0.0852	0.0951	0.0962
Sum	1	1	1	1	1	1	1	1	1
(CR, CI)	0.07	0.07	0.09	0.06	0.09	0.08	0.05	0.08	0.07

in Figures 7, 8 and 9, most of the current BSS stations are located in suitable and strong suitable regions.

After the alternative BSS station and BL are determined, the next step is to rank these alternatives by using the TOPSIS method so as to create more reliable suggestions for new investments related to cycling. In order to make TOPSIS calculations, normalized pixel values in the related criterion for each alternative should be obtained. All tables are not shown because of the word limitations. However, Table 4 that shows the normalized pixel values of BSS station alternatives for S3 is presented and Table 5 lists the ranks of each alternative. Besides, Fig. 10 shows the calculated rankings of BSS station alternatives for all scenarios. By doing so, alternative BSS station locations can be evaluated according to their rankings. While first rankings are A16, A10, and A15 for scenarios, respectively, worst rankings are A32, A30, and A26 for scenarios, respectively. Moreover, the rankings of each alternative BL are obtained as detailed in Fig. 11. Thus, it is determined which alternative BL is the best and which one is the worst. This supports decision makers to build BL.

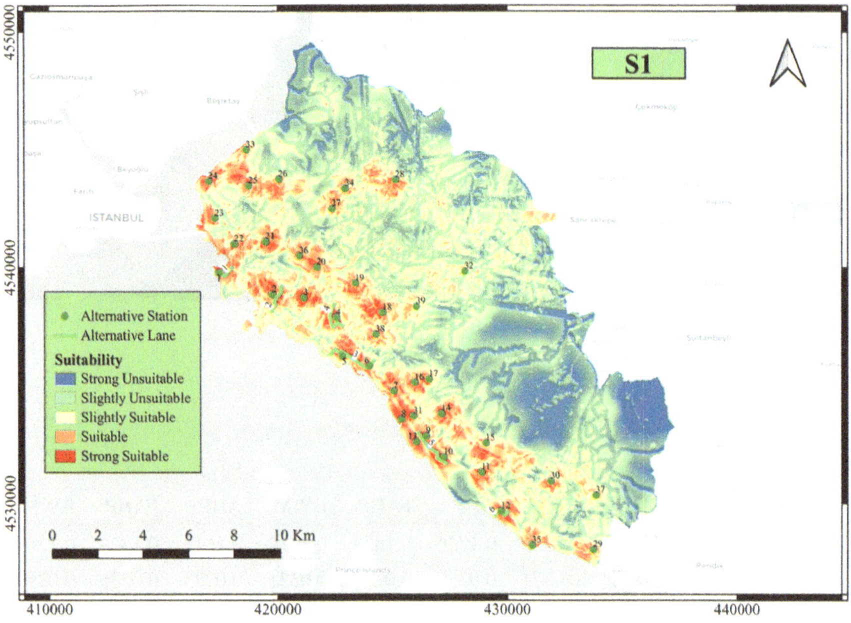

Fig. 7 Suitability map of S1

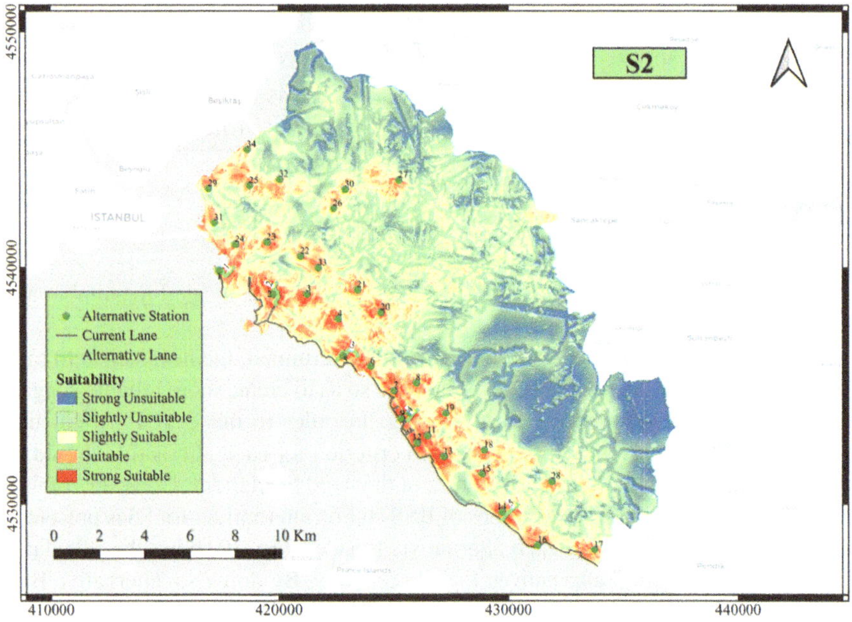

Fig. 8 Suitability map of S2

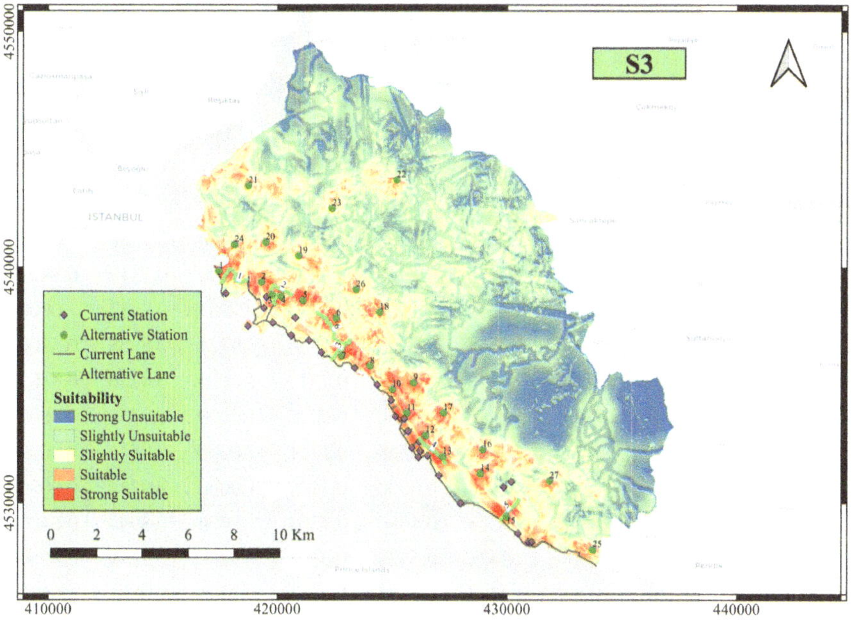

Fig. 9 Suitability map of S3

5 Conclusions

This study explores how to determine locations of the new BSS station and BL using open source GIS effectively. In this sense, we propose an approach that integrates MCDM and GIS in order to obtain results that are more efficient. This approach includes fuzzy logic modeling so as to represent the suitability of effective criteria more accurately in terms of spatiality. The significance of our study lies in finding the alternative locations of BSS station and BL at the same time by a hybrid approach. This would help the decision-makers for solutions to two different problems that affect each other. Moreover, three different MCDM methods are used for eliminating the disadvantages of every single method. The demand in cycling is delineated realistically since various effective criteria are taken into consideration. The findings of this study highlighted that open source GIS is a powerful tool for land use planning because it has capabilities to overcome different spatial analysis problems with the help of broad user and developer community. This enables us to conduct more efficient spatial decision analyses thanks to having the ability to be used without the need for any additional budget and reused by various analysts. Generated fuzzification model can be reused by researchers and applicants for different location selection problems since it allows essential fuzzification of vector data and is shared with the public. The other strong point of this study is that the proposed approach can be utilized for different study areas independently from containing BSS station and/or BL which was investigated by applying several scenarios. The results of scenarios

Table 4 Normalized pixel values

	C1	C2	C3	C4	C5	C6	C7	C8	C9
1	0.9998	0.8479	0.9600	0.9185	0.0000	0.4963	1.0000	1.0000	0.9773
2	0.6184	0.0000	1.0000	0.9998	0.0000	0.7687	1.0000	0.8787	0.8080
3	0.9107	0.0000	1.0000	1.0000	0.0000	0.7863	0.8574	0.9604	0.0000
4	0.9839	0.0000	0.9817	0.8479	0.0000	0.8271	1.0000	0.9426	0.4080
5	0.9768	0.0000	0.8200	0.9523	0.0000	0.9048	1.0000	0.2928	1.0000
6	0.7640	0.0000	0.8700	0.8479	0.9920	0.9420	1.0000	0.1119	0.6427
7	0.9431	0.0000	1.0000	0.8241	0.6069	0.7782	0.9983	0.9687	1.0000
8	0.9423	0.0000	1.0000	0.9634	0.0000	0.7776	0.9724	0.8958	0.6800
9	0.9478	0.6628	0.9000	0.9478	0.5992	0.7317	1.0000	0.4276	0.8387
10	0.9768	0.0000	1.0000	0.9988	0.0000	0.6636	1.0000	0.9758	0.0000
11	0.9702	0.0000	0.9817	0.9984	0.8568	0.5311	1.0000	0.9846	0.0000
12	0.9752	0.7000	0.8967	0.9608	0.0000	0.4199	1.0000	0.9344	0.0000
13	0.9966	0.0000	0.9300	0.9278	0.4080	0.3304	1.0000	0.8997	0.6400
14	0.7353	0.0000	1.0000	0.9431	0.6628	0.3535	0.9939	0.2279	0.8953
15	0.7928	0.0000	1.0000	0.9594	0.9313	0.2654	0.7101	0.9874	1.0000
16	0.8213	0.2439	0.8000	0.9988	0.3947	0.4438	0.8933	0.0000	0.4140
17	0.9423	0.1444	0.8800	0.9373	0.7625	0.5706	0.9732	0.0000	0.5580
18	1.0000	0.0000	0.8800	0.9423	0.3408	0.9873	0.9452	0.0000	0.0000
19	0.6948	0.1157	1.0000	0.9107	0.5638	0.9644	1.0000	0.0000	0.1547
20	0.9185	0.0000	1.0000	0.9752	0.8030	0.8489	0.9109	0.0000	0.0867
21	0.8335	0.6718	1.0000	0.9373	0.2227	0.6538	1.0000	0.0000	0.0000
22	0.8841	0.0000	1.0000	0.9702	0.0000	0.9676	0.9695	0.0000	0.0000
23	0.8375	0.0000	1.0000	0.9025	0.4473	0.9953	1.0000	0.0000	0.0000
24	0.7000	0.8852	1.0000	0.9025	0.1205	0.6427	1.0000	0.2518	0.2433
25	0.7640	0.0000	1.0000	0.8841	0.0000	0.1243	0.9965	0.8958	0.0000
26	0.9966	0.1920	0.2000	0.9768	0.0000	0.9985	1.0000	0.0000	0.0000
27	0.6628	0.0000	0.9600	0.9478	0.0000	0.3580	0.8947	0.0000	0.5460

are crucial since there are many cities without any BSS and BL currently. Urban and transportation planners can produce designments that ensure more livable and green cities. Further studies can focus on creating a web-based decision-making system by using open source tools in order to provide an easier solution without the need for any desktop software [71]. In spite of the fact that there are limitations from not having any questionnaire surveying for composing pairwise comparison matrices, we believe our work could be a framework for using open source GIS supported decision making with fuzzy logic in order to solve complex spatial problems that interconnect with urban areas.

Table 5 Ranks of the alternative BSS station for S3

	S_i+	S_i-	C_i	Rank
1	0.028466	0.049915	0.636829	4
2	0.033484	0.04319	0.563297	8
3	0.044278	0.036868	0.454343	16
4	0.036813	0.038589	0.511775	13
5	0.0364	0.040521	0.526788	11
6	0.031957	0.040572	0.559394	10
7	0.019689	0.051281	0.722571	2
8	0.032375	0.041873	0.563966	7
9	0.020634	0.042437	0.672842	3
10	0.044105	0.038102	0.463489	14
11	0.034878	0.0446	0.561162	9
12	0.042759	0.036496	0.460489	15
13	0.025328	0.041683	0.622034	5
14	0.028901	0.041597	0.59004	6
15	0.01864	0.053947	0.743204	1
16	0.03945	0.026765	0.404217	21
17	0.033859	0.035916	0.514745	12
18	0.047607	0.026739	0.359652	25
19	0.04299	0.029534	0.407229	20
20	0.042379	0.03435	0.447678	17
21	0.047653	0.027186	0.363258	24
22	0.05188	0.026389	0.33716	26
23	0.046929	0.028262	0.375868	23
24	0.041373	0.028774	0.410194	19
25	0.046155	0.033351	0.419474	18
26	0.055066	0.019519	0.261707	27
27	0.04534	0.027614	0.378515	22

S_i+: The separation from the negative ideal solution. S_i-: The separation of each alternative from the ideal solution. C_i: The relative closeness of the alternative

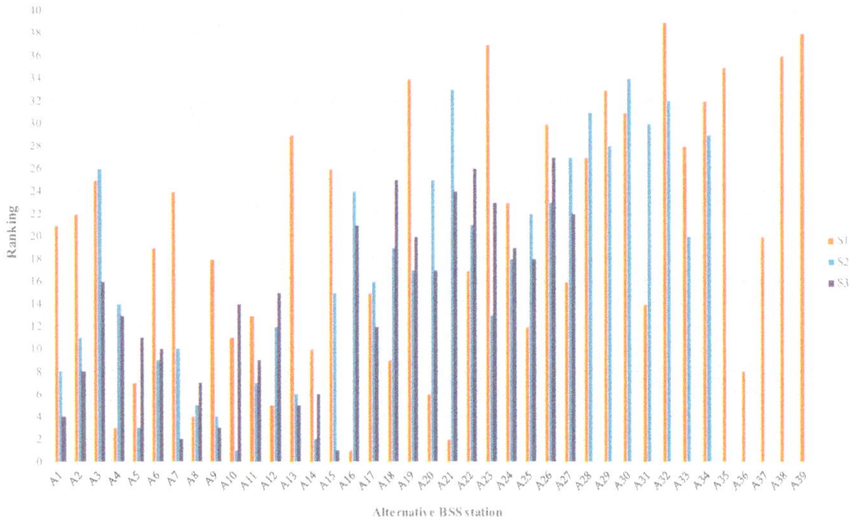

Fig. 10 Ranks of each alternative BSS station for all scenarios

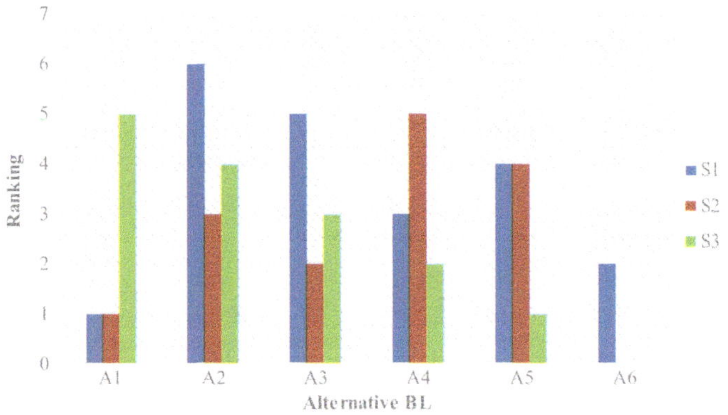

Fig. 11 Ranks of each alternative BL for all scenarios

References

1. Chen, Y., Bouferguene, A., Li, H.X., et al.: Spatial gaps in urban public transport supply and demand from the perspective of sustainability. J. Clean. Prod. **195**, 1237–1248 (2018). https://doi.org/10.1016/J.JCLEPRO.2018.06.021
2. Jain, D., Tiwari, G.: How the present would have looked like? Impact of non-motorized transport and public transport infrastructure on travel behavior, energy consumption and CO_2 emissions—Delhi, Pune and Patna. Sustain. Cities Soc. **22**, 1–10 (2016). https://doi.org/10.1016/J.SCS.2016.01.001

3. Israel Schwarzlose, A.A., Mjelde, J.W., Dudensing, R.M., et al.: Willingness to pay for public transportation options for improving the quality of life of the rural elderly. Transp. Res. Part A Policy Pract. **61**, 1–14 (2014). https://doi.org/10.1016/J.TRA.2013.12.009
4. Pucher, J., Peng, Z., Mittal, N., et al.: Urban transport trends and policies in China and India: impacts of rapid economic growth. Transp. Rev. **27**, 379–410 (2007). https://doi.org/10.1080/01441640601089988
5. Guler, D., Yomralioglu, T.: GIS and fuzzy AHP based area selection for electric vehicle charging stations. Int. Arch. Photogramm. Remote Sens. Spat. Inf. Sci. ISPRS Arch. 249–252 (2018). https://doi.org/10.5194/isprs-archives-XLII-4-249-2018
6. Goldman, T., Gorham, R.: Sustainable urban transport: four innovative directions. Technol. Soc. **28**, 261–273 (2006). https://doi.org/10.1016/J.TECHSOC.2005.10.007
7. Jones, L.R., Cherry, C.R., Vu, T.A., Nguyen, Q.N.: The effect of incentives and technology on the adoption of electric motorcycles: a stated choice experiment in Vietnam. Transp. Res. Part A Policy Pract. **57**, 1–11 (2013). https://doi.org/10.1016/J.TRA.2013.09.003
8. Lin, J.J., Lin, C.T., Feng, C.M.: Locating rental stations and bikeways in a public bike system. Transp. Plann. Technol. **41**, 402–420 (2018). https://doi.org/10.1080/03081060.2018.1453915
9. Fishman, E.: Bikeshare: a review of recent literature. Transp. Rev. **36**, 92–113 (2016). https://doi.org/10.1080/01441647.2015.1033036
10. Si, H., Shi, J., Wu, G., et al.: Mapping the bike sharing research published from 2010 to 2018: a scientometric review. J. Clean. Prod. **213**, 415–427 (2019). https://doi.org/10.1016/J.JCLEPRO.2018.12.157
11. Lin, J.-R., Yang, T.-H.: Strategic design of public bicycle sharing systems with service level constraints. Transp. Res. Part E Logist. Transp. Rev. **47**, 284–294 (2011). https://doi.org/10.1016/j.tre.2010.09.004
12. Conrow, L., Murray, A.T., Fischer, H.A.: An optimization approach for equitable bicycle share station siting. J. Transp. Geogr. **69**, 163–170 (2018). https://doi.org/10.1016/j.jtrangeo.2018.04.023
13. Abolhassani, L., Afghari, A.P., Borzadaran, H.M.: Public preferences towards bicycle sharing system in developing countries: the case of Mashhad, Iran. Sustain. Cities Soc. **44**, 763–773 (2019). https://doi.org/10.1016/J.SCS.2018.10.032
14. Kondo, M.C., Morrison, C., Guerra, E., et al.: Where do bike lanes work best? A Bayesian spatial model of bicycle lanes and bicycle crashes. Saf. Sci. **103**, 225–233 (2018). https://doi.org/10.1016/J.SSCI.2017.12.002
15. Lee, S.E., Simons-Morton, B.G., Klauer, S.E., et al.: Naturalistic assessment of novice teenage crash experience. Accid. Anal. Prev. **43**, 1472–1479 (2011). https://doi.org/10.1016/J.AAP.2011.02.026
16. Faghih-Imani, A., Anowar, S., Miller, E.J., Eluru, N.: Hail a cab or ride a bike? A travel time comparison of taxi and bicycle-sharing systems in New York City. Transp. Res. Part A Policy Pract. **101**, 11–21 (2017). https://doi.org/10.1016/J.TRA.2017.05.006
17. Cai, S., Long, X., Li, L., et al.: Determinants of intention and behavior of low carbon commuting through bicycle-sharing in China. J. Clean. Prod. **212**, 602–609 (2019). https://doi.org/10.1016/J.JCLEPRO.2018.12.072
18. Jain, T., Wang, X., Rose, G., Johnson, M.: Does the role of a bicycle share system in a city change over time? A longitudinal analysis of casual users and long-term subscribers. J. Transp. Geogr. **71**, 45–57 (2018). https://doi.org/10.1016/J.JTRANGEO.2018.06.023
19. Hyland, M., Hong, Z., de Farias Pinto, H.K.R., Chen, Y.: Hybrid cluster-regression approach to model bikeshare station usage. Transp. Res. Part A Policy Pract. **115**, 71–89 (2018). https://doi.org/10.1016/J.TRA.2017.11.009
20. Heinen, E., Kamruzzaman, M., Turrell, G.: The public bicycle-sharing scheme in Brisbane, Australia: evaluating the influence of its introduction on changes in time spent cycling amongst a middle- and older-age population. J. Transp. Health **10**, 56–73 (2018). https://doi.org/10.1016/J.JTH.2018.07.003
21. Van Cauwenberg, J., Clarys, P., De Bourdeaudhuij, I., et al.: Environmental influences on older adults' transportation cycling experiences: a study using bike-along interviews. Landsc. Urban Plan. **169**, 37–46 (2018). https://doi.org/10.1016/J.LANDURBPLAN.2017.08.003

22. Wu, J., Wang, L., Li, W.: Usage patterns and impact factors of public bicycle systems: comparison between city center and suburban district in Shenzhen. J. Urban Plan. Dev. **144**, 4018027 (2018). https://doi.org/10.1061/(ASCE)UP.1943-5444.0000471
23. Sun, Y., Mobasheri, A., Hu, X., et al.: Investigating impacts of environmental factors on the cycling behavior of bicycle-sharing users. Sustainability **9**, 1060 (2017). https://doi.org/10.3390/su9061060
24. Yuan, M., Zhang, Q., Wang, B., et al.: A mixed integer linear programming model for optimal planning of bicycle sharing systems: a case study in Beijing. Sustain. Cities Soc. **47**, 101515 (2019). https://doi.org/10.1016/J.SCS.2019.101515
25. Cheng, Y.H., Lin, Y.C.: Expanding the effect of metro station service coverage by incorporating a public bicycle sharing system. Int. J. Sustain. Transp. **12**, 241–252 (2018). https://doi.org/10.1080/15568318.2017.1347219
26. Griffin, G.P., Jiao, J.: Crowdsourcing bike share station locations. J. Am. Plan. Assoc. **85**, 35–48 (2019). https://doi.org/10.1080/01944363.2018.1476174
27. Wang, L., Li, C., Chen, M.Z.Q., et al.: Connectivity-based accessibility for public bicycle sharing systems. IEEE Trans. Autom. Sci. Eng. **15**, 1521–1532 (2018). https://doi.org/10.1109/TASE.2018.2868471
28. Boettge, B., Hall, D., Crawford, T., et al.: Assessing the bicycle network in St. Louis: a place based user-centered approach. Sustainability **9**, 241 (2017). https://doi.org/10.3390/su9020241
29. Mooney, S.J., Hosford, K., Howe, B., et al.: Freedom from the station: spatial equity in access to dockless bike share. J. Transp. Geogr. **74**, 91–96 (2019). https://doi.org/10.1016/J.JTRANGEO.2018.11.009
30. Hosford, K., Winters, M.: Who are public bicycle share programs serving? An evaluation of the equity of spatial access to bicycle share service areas in Canadian cities. Transp. Res. Rec. J. Transp. Res. Board **2672**, 42–50 (2018). https://doi.org/10.1177/0361198118783107
31. Ferenchak, N.N., Marshall, W.E.: Suppressed child pedestrian and bicycle trips as an indicator of safety: adopting a proactive safety approach. Transp. Res. Part A Policy Pract. **124**, 128–144 (2019). https://doi.org/10.1016/J.TRA.2019.03.010
32. Saplıoğlu, M., Aydın, M.M.: Choosing safe and suitable bicycle routes to integrate cycling and public transport systems. J. Transp. Health (2018). https://doi.org/10.1016/j.jth.2018.05.011
33. Kaygisiz, Ö., Hauger, G.: Network-based point pattern analysis of bicycle accidents to improve cyclist safety. Transp. Res. Rec. **2659**, 106–116 (2017). https://doi.org/10.3141/2659-12
34. Lin, J.J., Yu, C.J.: A bikeway network design model for urban areas. Transportation **40**, 45–68 (2013). https://doi.org/10.1007/s11116-012-9409-6
35. Kent, M., Karner, A.: Prioritizing low-stress and equitable bicycle networks using neighborhood-based accessibility measures. Int. J. Sustain. Transp. **13**, 100–110 (2019). https://doi.org/10.1080/15568318.2018.1443177
36. Veillette, M.P., Grisé, E., El-Geneidy, A.: Park ‘n’ roll: identifying and prioritizing locations for new bicycle parking in Québec City, Canada. Transp. Res. Rec. J. Transp. Res. Board **2672**, 73–82 (2018). https://doi.org/10.1177/0361198118776522
37. Larsen, J., Patterson, Z., El-Geneidy, A.: Build it. But where? The use of geographic information systems in identifying locations for new cycling infrastructure. Int. J. Sustain. Transp. **7**, 299–317 (2013). https://doi.org/10.1080/15568318.2011.631098
38. Rybarczyk, G., Wu, C.: Bicycle facility planning using GIS and multi-criteria decision analysis. Appl. Geogr. **30**, 282–293 (2010). https://doi.org/10.1016/j.apgeog.2009.08.005
39. García-Palomares, J.C., Gutiérrez, J., Latorre, M.: Optimizing the location of stations in bike-sharing programs: a GIS approach. Appl. Geogr. **35**, 235–246 (2012). https://doi.org/10.1016/j.apgeog.2012.07.002
40. Terh, S.H., Cao, K.: GIS-MCDA based cycling paths planning: a case study in Singapore. Appl. Geogr. **94**, 107–118 (2018). https://doi.org/10.1016/j.apgeog.2018.03.007
41. Kabak, M., Erbaş, M., Çetinkaya, C., Özceylan, E.: A GIS-based MCDM approach for the evaluation of bike-share stations. J. Clean. Prod. **201**, 49–60 (2018). https://doi.org/10.1016/j.jclepro.2018.08.033

42. Sun, Y., Mobasheri, A.: Utilizing crowdsourced data for studies of cycling and air pollution exposure: a case study using strava data. Int. J. Environ. Res. Public Health **14**, 274 (2017). https://doi.org/10.3390/ijerph14030274
43. Sun, Y., Moshfeghi, Y., Liu, Z.: Exploiting crowdsourced geographic information and GIS for assessment of air pollution exposure during active travel. J. Transp. Health **6**, 93–104 (2017). https://doi.org/10.1016/j.jth.2017.06.004
44. Sun, Y., Du, Y., Wang, Y., Zhuang, L.: Examining associations of environmental characteristics with recreational cycling behaviour by street-level strava data. Int. J. Environ. Res. Public Health **14**, 644 (2017). https://doi.org/10.3390/ijerph14060644
45. Conrow, L., Wentz, E., Nelson, T., Pettit, C.: Comparing spatial patterns of crowdsourced and conventional bicycling datasets. Appl. Geogr. **92**, 21–30 (2018). https://doi.org/10.1016/j.apgeog.2018.01.009
46. Norman, P., Pickering, C.M., Castley, G.: What can volunteered geographic information tell us about the different ways mountain bikers, runners and walkers use urban reserves? Landsc. Urban Plan. **185**, 180–190 (2019). https://doi.org/10.1016/j.landurbplan.2019.02.015
47. Orellana, D., Guerrero, M.L.: Exploring the influence of road network structure on the spatial behaviour of cyclists using crowdsourced data. Environ. Plan. B Urban Anal. City Sci. **46**, 1314–1330 (2019). https://doi.org/10.1177/2399808319863810
48. Giuffrida, L.P., Inturri, I.: Mapping with stakeholders: an overview of public participatory GIS and VGI in transport decision-making. ISPRS Int. J. Geo-Inf. **8**, 198 (2019). https://doi.org/10.3390/ijgi8040198
49. Güler, D., Yomralıoğlu, T.: Alternative suitable landfill site selection using analytic hierarchy process and geographic information systems: a case study in Istanbul. Environ. Earth Sci. 76 (2017). https://doi.org/10.1007/s12665-017-7039-1
50. Zadeh, L.A.: Fuzzy sets. Inf. Control **8**, 338–353 (1965). https://doi.org/10.1016/S0019-9958(65)90241-X
51. Kaufmann, A., Gupta, M.M.: Fuzzy Mathematical Models in Engineering and Management Science. Elsevier Science Inc. (1988)
52. Zimmermann, H.J.: Fuzzy Set Theory—And Its Applications. Springer, Dordrecht (2011)
53. Malczewski, J., Rinner, C.: Dealing with uncertainties. In: Malczewski, J., Rinner, C. (eds.) Multicriteria Decision Analysis in Geographic Information Science, pp. 191–221. Springer Berlin Heidelberg, Berlin, Heidelberg (2015)
54. Zeng, T.Q., Zhou, Q.: Optimal spatial decision making using GIS: a prototype of a real state geographical information system (REGIS). Int. J. Geogr. Inf. Sci. (2001). https://doi.org/10.1080/136588101300304034
55. Lin, C.T., Lin, J.K.: Fuzzy-GIS approach for applying the AHP multi-criteria decision-making model to evaluate real estate purchases. J. Test. Eval. (2013). https://doi.org/10.1520/jte20120030
56. Saaty, T.L., Vargas, L.G.: How to make a decision. In: Models, Methods, Concepts & Applications of the Analytic Hierarchy Process, pp. 1–21. Springer US, Boston, MA (2012)
57. Şen, C.G., Çınar, G.: Evaluation and pre-allocation of operators with multiple skills: a combined fuzzy AHP and max–min approach. Expert Syst. Appl. **37**, 2043–2053 (2010). https://doi.org/10.1016/J.ESWA.2009.06.075
58. Rezaei, J.: Best-worst multi-criteria decision-making method. Omega **53**, 49–57 (2015). https://doi.org/10.1016/j.omega.2014.11.009
59. Rezaei, J.: Best-worst multi-criteria decision-making method: some properties and a linear model. Omega **64**, 126–130 (2016). https://doi.org/10.1016/j.omega.2015.12.001
60. Hwang, C.L., Yoon, K.: Multiple Attribute Decision Making: Methods and Applications: A State-of-the-Art Survey. Springer-Verlag, New York (1981)
61. Neteler, M., Bowman, M.H., Landa, M., Metz, M.: GRASS GIS: a multi-purpose open source GIS. Environ. Model. Softw. **31**, 124–130 (2012). https://doi.org/10.1016/J.ENVSOFT.2011.11.014
62. Arias de Reyna, M., Simoes, J.: Empowering citizen science through free and open source GIS. Open Geospat. Data Softw. Stand. 1 (2016). https://doi.org/10.1186/s40965-016-0008-x

63. Grizonnet, M., Michel, J., Poughon, V., et al.: Orfeo ToolBox: open source processing of remote sensing images. Open Geospat. Data Softw. Stand. **2**, 15 (2017). https://doi.org/10.1186/s40965-017-0031-6
64. Ledoux, H.: val3dity: validation of 3D GIS primitives according to the international standards. Open Geospat. Data Softw. Stand. 3 (2018). https://doi.org/10.1186/s40965-018-0043-x
65. Steiniger, S., Hunter, A.J.S.: The 2012 free and open source GIS software map—a guide to facilitate research, development, and adoption. Comput. Environ. Urban Syst. **39**, 136–150 (2013)
66. Swain, N.R., Latu, K., Christensen, S.D., et al.: A review of open source software solutions for developing water resources web applications. Environ. Model. Softw. **67**, 108–117 (2015)
67. Dile, Y.T., Daggupati, P., George, C., et al.: Introducing a new open source GIS user interface for the SWAT model. Environ. Model. Softw. **85**, 129–138 (2016). https://doi.org/10.1016/J.ENVSOFT.2016.08.004
68. Petrasova, A., Petras, V., Harmon, B., Mitasova, H.: Tangible Modeling with Open Source GIS. Springer, Cham (2015)
69. Steiniger, S., Bocher, E.: An overview on current free and open source desktop GIS developments. Int. J. Geogr. Inf. Sci. **23**, 1345–1370 (2009). https://doi.org/10.1080/13658810802634956
70. Milakis, D., Athanasopoulos, K.: What about people in cycle network planning? Applying participative multicriteria GIS analysis in the case of the Athens metropolitan cycle network. J. Transp. Geogr. **35**, 120–129 (2014). https://doi.org/10.1016/j.jtrangeo.2014.01.009
71. Mete, M.O., Guler, D., Yomralioglu, T.: Development of 3D web GIS application with open source library. Selcuk Univ. J. Eng. Sci. Technol. **6**, 818–824 (2018). https://doi.org/10.15317/Scitech.2018.171

Spatial Query Performance Analyses on a Big Taxi Trip Origin–Destination Dataset

Berk Anbaroğlu

Abstract Database management systems (DBMSs) are widely used to store, query and visualise large amounts of spatial data. Consequently, teaching DBMSs at both undergraduate and graduate level is a core component within most of the geoinformatics related departments' curricula. Traditional teaching material mainly covers relational DBMS, due to their strong linkage with GIS. However, due to the emergence of web-based systems, employers also require prospective graduates to have experience on non-relational DBMS (NoSQL). This chapter develops an open-source library to analyse the query performance of two renowned DBMS: PostgreSQL/PostGIS and MongoDB, a relational and NoSQL DBMS respectively. An open-source Python library is developed to facilitate systematic performance analyses between these DBMSs. The experiments are carried out on New York City's openly available taxi trip origin–destination dataset. The performance of two spatial queries (k-nearest neighbour and point-in-polygon) are investigated in terms of run-time and spatial accuracy. The results indicate the superiority of MongoDB. It outperformed Postgres in terms of run-time in both of the investigated queries. In addition, it is more accurate in terms of detecting k-nearest neighbours. The developed open-source library is utilised to investigate journey time variations between two airports of New York City, which demonstrates its effectiveness in terms of teaching DBMS or GIS modules.

Keywords Database management system · Spatial query performance · Open source · GIS

1 Introduction

The wide use of sensors and mobile devices have enabled the generation of unprecedented volumes of spatial data. In order to effectively process such large amounts of

B. Anbaroğlu (✉)
Department of Geomatics Engineering, Hacettepe University, Ankara, Turkey
e-mail: banbar@hacettepe.edu.tr

A. Mobasheri (ed.), *Open Source Geospatial Science for Urban Studies*, Lecture Notes in Intelligent Transportation and Infrastructure,
https://doi.org/10.1007/978-3-030-58232-6_3

data, researchers and analysts often store their spatial data in a DataBase Management System (DBMS) prior to visualizing it in a Geographical Information System (GIS). However, it is not a straightforward task to choose the correct DBMS, due to myriad of alternatives. There are more than 350 DBMSs as reported in 'db-engines.com' web-site, which compares the popularity of different DBMSs. The top four of the list are represented by relational databases. The success of relational databases are due to their natural linkage with Structured Query Language (SQL), which allows to query data in a straightforward manner. This feature of relational DBMS played a key role in the proliferation of GIS. Specifically, by relying on relational DBMS, it is possible to link different spatial layers through common attributes or simply geographical location [1, 2]. Consequently, teaching relational DBMS constitutes a major component of most DBMS or GIS modules at both undergraduate and graduate programs.

On the other hand, relational DBMSs might fail to incorporate the dynamic structure of web-based systems. The emergence of big data and cloud computing has led researchers and practitioners to investigate the effectiveness of NoSQL (Not Only SQL) databases, which are also referred as schema-less databases. There are several advantages of relying on NoSQL databases with respect to relational databases. First, high concurrent reading and writing with low latency could be achieved. In this way, many users can interact with the database in real-time. Second, they provide an infrastructure suitable for storing large amounts of heterogeneous data. Third, rapid uninterrupted service could be provided as NoSQL databases are horizontally scalable [3]. The pervasive use of NoSQL DBMSs are due to the success of Web 2.0 companies, as these companies develop and maintain such systems (e.g. Bigtable and DynamoDB are developed by Google and Amazon respectively).

In order to deliver a DBMS or GIS module that meets industry standards, it is therefore necessary for educators to provide hands-on experiments to students on not only relational DBMSs but also NoSQL DBMSs. In this way, students could test the different aspects of DBMSs ranging from the ease of use to the way in which queries are processed. Even though it would not be feasible to test the entire set of available DBMSs within an academic year, it is still important to provide the opportunity to test at least two different DBMSs. The openly available large spatial datasets are invaluable assets in this process. For example, New York City Taxi and Limousine Commission (TLC) openly distributes the Origin–Destination of taxi trips since 2009 [4]. Such large datasets, coupled with additional spatial layers including taxi zones or road network would provide the necessary means for educators to deliver different assignments.

The aim of this chapter is to develop an open-source library for analysing the query performance of two commonly used DBMSs: PostgreSQL (Postgres hereafter) and MongoDB. Postgres is a renowned example of a relational DBMS. On the other hand, MongoDB is a renowned example of NoSQL. Both of these DBMSs support spatial data types and spatial indexing. Therefore, both are suitable for querying large spatial datasets. The experiments are carried out on New York City's openly available taxi dataset. Investigated spatial queries include finding the k-Nearest Neighbour (KNN) of a point, and point-in-polygon analyses. In addition the developed library includes some useful functions for processing the databases. These functions are utilised to investigate the journey time variation between John F. Kennedy (JFK) and LaGuardia (LGA) airports. Consequently, this chapter contributes to the literature by first developing an open-source library that could be used within DBMS or GIS modules [5]. In addition, it analyses the query performance of Postgres and MongoDB, both of which are widely used for spatial analysis.

The organization of this chapter is as follows. Section 2 discusses the relevant literature on relational and NoSQL databases as well as urban traffic data. Section 3 describes the developed open-source library and compares both the query performance of Postgres and MongoDB on a large scale and Origin–Destination (OD) dataset in various aspects. Finally, Sect. 4 concludes this chapter by discussing the findings of this research and elaborates on the future research directions.

2 Literature Review

This section first reviews the relevant literature on relational and NoSQL DBMS. Specifically, different types of NoSQL DBMSs and the metrics to compare them are introduced. Second, the use-cases of origin–destination datasets are described, since such an exemplar dataset is investigated in the analyses.

2.1 Relational and NoSQL DBMSs

Traditional way to store and manage data is to rely on relational DBMSs such as MySQL, Oracle or Postgres. The main advantage of using a relational DBMS is the Atomicity–Consistency–Isolation–Duration (ACID) feature. In this way, various desired features of modern GIS including data integrity, database security and efficient querying could be achieved. Relational DBMSs are still commonly used, but they require a pre-defined schema indicating the relationships between the entities (i.e. tables). Defining and updating such a schema might be difficult in a dynamic setting such as the web. Therefore, researchers developed alternative DBMSs, which do not require a pre-defined schema, also known as Not-Only SQL (NoSQL) DBMSs. Different types of NoSQL DBMSs are: (i) 'Document Store' (e.g. MongoDB and CouchDB), (ii) 'Wide Column Store/Column Families' (e.g. Hadoop/HBase and Cassandra), and (iii) 'Graph Databases' (e.g. Neo4J and OrientDB) [6].

Comparative studies between different DBMSs have already attracted the attention of researchers and analysts. The comparisons between different DBMSs are based on various issues ranging from read-write-update latency [3, 7] to the use of data structures for indexing [8] and to database security [9]. It is not surprising to observe that different DBMSs have different advantages and limitations. Consequently, [10] stated that 'a definite decision for a database management system is not possible' and proposed a decision-support framework to balance the requirements

Table 1 DBMSs that could handle spatial data

Relational DBMS	NoSQL DBMS
Postgres + PostGIS extension	MongoDB
Oracle spatial	CouchDB + GeoCouch extension
MySQL	Neo4j
SQL server	BigTable
	HBase
	Cassandra

of query optimization and application-performance estimation. Nevertheless, only a few DBMSs support spatial queries and they are reported in Table 1.

Recent research evidence suggests that NoSQL DBMSs might perform better compared to relational DBMSs in terms of spatial queries. For instance, [11] investigated the effectiveness of HBase, which has a row-key design for indexing, to process GPS trajectories so that they match with the correct road (i.e. map-matching). They analysed historic spatial data without frequent insertions or updates. The results suggests that HBase is suitable for the task as it substantially improved the processing speed with respect to MySQL. Another research compared Postgres, MySQL and Oracle (relational DBMSs) with Jena and Sesame (NoSQL DBMSs) [12]. Their findings indicate that Postgres and Oracle are better in terms of point-in-polygon query execution time and loading data respectively. However, only circles are considered within the point-in-polygon analysis.

The natural linkage between Postgres and QGIS makes Postgres a reasonable choice to store spatial data. On the other hand, MongoDB provides inherent spatial indexing. As both of these DBMSs are open-source, they possess a considerable user-base, indicating that technical problems could be solved quickly. Consequently, researchers focused on these two DBMSs while analysing spatial data. For example, [13] investigated the web service request performance of point-in-polygon analysis on Postgres and MongoDB, which is an important metric regarding the Infrastructure for Spatial Information in Europe (INSPIRE) directive [14]. Even though there is no decisive advantage of one to another, complex geographic data types such as polygons are better handled by Postgres. On the other hand, MongoDB serve point data faster regardless the database size. Another study investigating line intersection and point-in-polygon analyses adds further support to this finding by stating that 'NoSQL databases may be better suited for simultaneous multiple-user query systems including Web-GIS and mobile-GIS' [15]. A more recent research conducted by [16] compared Postgres and MongoDB in terms of KNN execution times and spatial accuracy and found out that MongoDB is better on both aspects.

2.2 *Urban Traffic Data*

Traffic operation centers and companies are collecting large amounts of urban traffic data on a daily basis and in real-time. Effective management of such big data is gaining importance to provide a better service [17, 18]. Some cities distribute their data through open standards like General Transit Feed Specification (GTFS) or simply spreadsheets to foster collaboration and advance economics. Online visualization and distribution of large datasets has already emerged as a research agenda [19]. Relying on open standards and open data are key requirements to fostering collaboration between researchers and practitioners. In this way, there is an unprecedented opportunity to have a better understanding of urban dynamics [20].

Origin–Destination (OD) data are useful to understand different aspects of an urban transportation system especially when OD data are collected in large quantities, on a regular basis, and across an urban road network. For instance, [21] estimated urban link journey times by optimizing the number of license plate scanning devices and their locations. Barann et al. [22] proposed a taxi-sharing framework using archived OD data, which had proven to be cost-effective in terms of both the passengers and the taxi driver. Zhu et al. [23] evaluated the resilience of transportation systems with respect to natural disasters by utilizing millions of taxi OD data. These studies demonstrate the use cases of large-scale OD datasets and address the necessity to manage them effectively. Therefore, performance wise studies also attract the attention of researchers [16, 24].

The effective management of large OD datasets often requires the use of a DBMS. Indeed, most transportation research activities dealing with large amounts of data rely on DBMSs including, but not limited to, ridesharing planning [25], trajectory pattern mining [26] and incident management [27]. Even though some studies overly mention the word 'database', it is not clear which DBMS they have relied on [28]. It is also a common practice in transportation science to rely on a single DMBS such as MySQL [29]. Considering that many day-to-day operational tasks to manage an urban road network efficiently would require the effective use of a DBMS, it is important to assess the capabilities of alternatives.

3 Case Study: Taxi Trips of New York City

This chapter analyses the spatial query performance of Postgres and MongoDB and develops an open-source library accordingly [5]. Both of these DBMSs support (i) spatial data types, (ii) spatial indexing and (iii) they are both open-source projects and possess a considerable user-base. The experiments are carried out on the openly available taxi trip data provided by the Taxi and Limousine Commission (TLC) of New York City (NYC) [4].

3.1 Data Import

There are three main types of taxis in NYC: yellow, green and for-hire vehicle. Taxi data are available online for all these three types of taxis; however, only the yellow taxi dates back to 2009. Considering its historic availability and spatial coverage, this chapter relies on the yellow taxi data records. Each trip is described by 19 attributes including the pickup and dropoff location and time, trip distance, total fare and the number of passengers. The yellow taxi trips of 2015 are analysed, which constitutes of approximately 144 million taxi trips.

The taxi trips do not possess a unique ID. In order to provide a unique ID to each trip, data are first imported into Postgres. While creating a table, it is required to determine the suitable data types for each attribute. Therefore, a useful laboratory assignment would be investigating the size-on-disk effectiveness of two different tables. In the first table, all attributes would be set to 'text', which is straightforward to do but not meaningful. In the second table, all attributes would be assigned to their optimal data types. This could be achieved by inspecting on the metadata. Once the data are imported into both tables, it would be valuable to observe that the former not only wastes space, but also prohibit querying the data.

In the 'staging' table, the pickup and dropoff locations are stored as a real number. After enabling the PostGIS extension, the conversion from real latitude/longitude attributes to a single geometry attribute would be achieved to facilitate spatial analyses. It is possible to hold two geometries in one record; hence, both the pickup and dropoff geometries could be kept in PostGIS.

The second step is to import the data into MongoDB. In order to facilitate this, a function is written to convert Postgres records into suitable GeoJSON files. It is possible to have a timestamp without a time zone in Postgres. However, time stamps in MongoDB requires the date/time to be in ISO 8601 standard, which necessitates the use of time zones. For example, the date/time value '2015-01-10 20:33:41' in Postgres must be converted into '2015-01-10T20:33:41Z' in order to satisfy MongoDB's requirement. Consequently, the function 'rearrangeTimeFormat' is written, which could be accessed from the 'postgres2GeoJSON' script. The idea in this process is to partition the 'staging' table of Postgres based on the unique IDs of records, and then create a GeoJSON file by appending the records. Having achieved this, the GeoJSON files could be imported into MongoDB using Studio 3T, which is a proprietary software providing a Graphical User Interface (GUI) for MongoDB operations. While importing chunks of taxi trips into MongoDB one chunk failed to load, due to an unknown reason, where each chunk contained two million trips. The overall comparison of the two databases is illustrated in Table 2.

An overview of the study area is presented in Fig. 1 where the pickup locations of yellow taxi trips are illustrated for a randomly chosen day, 22 August 2015. This visualization reflects the spatial heterogeneity regarding the concentration of pickup locations. For instance, there were no pickups from the Staten Island, whereas the financial district Manhattan has very high pickup density. In addition, 263 taxi zones provided by TLC are also illustrated with a highlight on the airports: John F.

Table 2 Overall comparison of the databases

	Postgres	MongoDB
Version	9.3 PostGIS version: 2.2.3	4.1.6
GUI	pgAdmin III	Studio 3T
Database size	27 GB	21.9 GB
License	Open Source Initiative (OSI)	GNU AGPL v3.0
Number of trips	144,112,989	144,112,989

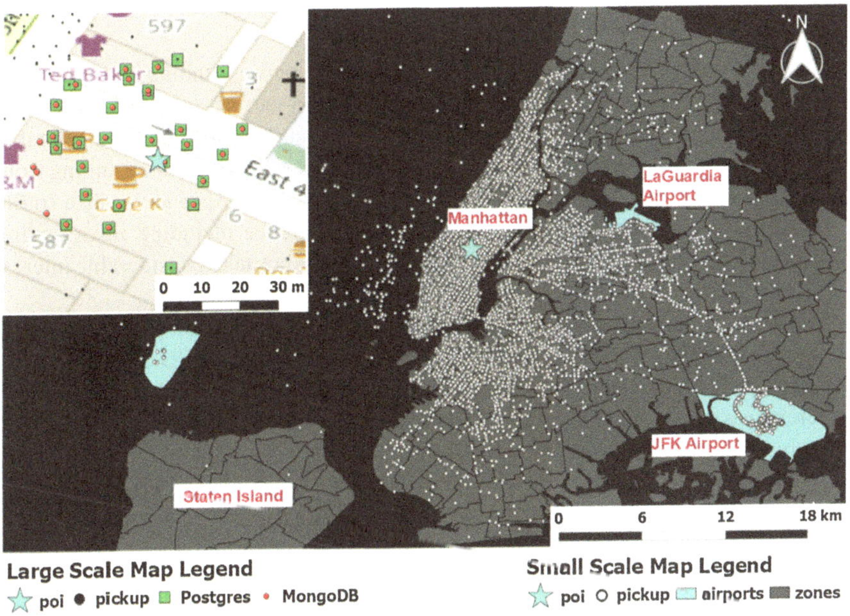

Fig. 1 The pickup locations on 22 August 2015 and the mismatch of KNN for Postgres and MongoDB

Kennedy on the south, LaGuardia on the north, and Newark on the west. The KNN of a randomly chosen point are also illustrated in Fig. 1, where it is clear that Postgres and MongoDB detected different sets of nearest neighbours of the same point. This issue would be discussed in more detail in the subsequent subsection.

The aim of this chapter is to develop an open-source library to facilitate the query performance analyses between Postgres and MongoDB in a systematic way [5]. However, such a comparative study could involve many experiments. Some previous studies have already analysed the performance of indexed versus non-indexed structures. As expected, these studies found out that indexing improves the overall execution time [30]. Therefore, prior to continuing the empirical investigation, an index is created on the location and time of pickup/dropoff locations. Physical design of databases is a complex task, involving the tuning of different parameters such as shared buffer and maintenance work memory [30]. In order to lower the administration costs, default parameter settings are used for Postgres and MongoDB [31]. As a result, pickup/dropoff locations are indexed using GIST and 2DSphere in Postgres and MongoDB respectively; and pickup/dropoff times are indexed using B-tree in both Postgres and MongoDB. The experiments are carried out on a computer running Windows 10 with 16 GB of RAM and 3.6 GHz of i7 CPU.

3.2 k Nearest Neighbour Analysis

A common task in GIS, as well as in other disciplines, is finding the k-Nearest Neighbour (KNN) of a query instance. The operation can be applied for different purposes including classification and clustering of data; hence, takes considerable attention from researchers [32]. Applications of this operation is diverse ranging from supervised lithology classification in geological mapping [33] to short-term traffic forecasting [34]. Consequently, performance-wise studies including investigating efficient ways to index spatial data [35] to empirical studies take considerable interest [16].

This subsection describes the experimental design and results of KNN query on Postgres and MongoDB. Random trips are chosen and their KNN are determined by relying on its coordinates. On Postgres, this query can be implemented in two different ways. First, the pair wise distances between all the points are calculated and then these distances are sorted to obtain the k nearest ones (Postgres-v1). Note that the '<–>' symbol calculates the distance between two points. Second, a subquery is included to select the location of the point-of-interest and then the distance from this point to the rest is determined (i.e. Postgres–v2). This approach is indeed what MongoDB relies on. In MongoDB, first the longitude and latitude of the point is determined (i.e. p.x and p.y) through a query and these values are then fed into the *$nearSphere* function. These query models are illustrated in Fig. 2.

The performance of the KNN query is evaluated in two different aspects: run time and spatial accuracy. The first approach in Postgres has empirically shown to be much slower compared to the second version [16]. Therefore, this chapter focused on the performance assessment of Postgres-v2 and MongoDB.

The KNN of thirty random pickup locations are determined. The average run time of these thirty queries with respect to different values of k are illustrated in Fig. 3. MongoDB, regardless the value of k, detected the neighbours of the query point almost instantly.

The empirical results suggest that the detected neighbours are different between Postgres and MongoDB. In other word, the neighbours detected by Postgres do not match with what MongoDB had detected. Spatial accuracy is assessed by relying on the Haversine distance. Assume that the maximum distance between the query point i and its most distant nearest neighbour is denoted with in metres. Amongst the randomly generated thirty points (i.e. $i \in [1, 2, \ldots, 30]$), Table 3 illustrates the

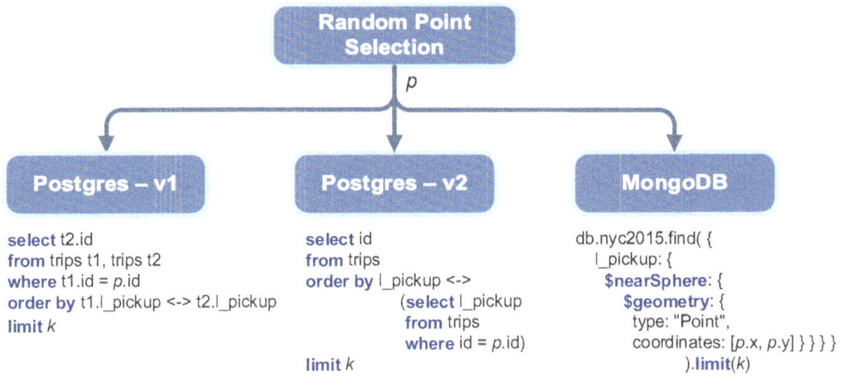

Fig. 2 Different KNN query models

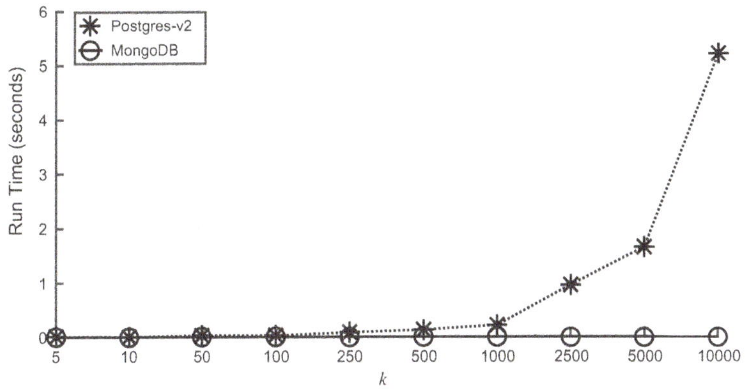

Fig. 3 Run times of Postgres-v2 and MongoDB regarding the KNN query

Table 3 Maximum Haversine distance between the query points and their most distant neighbour

k	5	10	50	100	250	500	1000	2500	5000	10,000
M	1.97	3.82	7.66	10.56	14.64	17.83	23.02	37.31	52.74	68.70
P	1.97	3.82	8.16	12.03	16.54	20.40	26.42	45.08	64.17	81.45

maximum for different values of *k*, where M and P denotes MongoDB and Postgres respectively.

The results indicate that MongoDB is not only much faster but also detects spatially accurate neighbours. The code to analyse for a single point and a single *k* value is available on GitHub [5]. For a more comprehensive experimentation, the available script (i.e. KNN.py) could be extended to include Postgres-v1. The results obtained here are in line with what had previously been published [16].

3.3 Point-In-Polygon Analysis

The aim of point-in-polygon (pip) analysis is to determine the polygon which a query point resides in. Having its root in computer graphics and computational geometry, several algorithms have been proposed to identify whether a point is inside a polygon or not including even–odd rule or winding number [36]. It is still on the research agenda of the researchers since some edge conditions or complex polygons might return erroneous outcomes or cause inefficiencies [37, 38]. The use cases of pip analysis is diverse, and includes the construction of an indoor air pollution monitoring system [39], understanding biodiversity [40] and autonomous navigation of robots [41].

This subsection describes the performance assessment of Postgres and MongoDB with respect to the pip query. First, the taxi zones provided by TLC as a shapefile is used, which contains 263 polygons. Each taxi zone is represented by a zone ID, name and borough. Taxi zones provided by TLC are in local geographical projection. QGIS is used to transform the zones into WGS84 so that spatial overlay between zones and taxi pickup/dropoff locations is satisfied. The taxi zones shapefile is imported into Postgres as a new table, and into MongoDB as an attachment to the existing collection.

The formalization of the queries are illustrated in Fig. 4. Postgres query joins two distinct tables (trips and zones) based on location and determines the polygon which the query point is in accordingly. On the other hand, MongoDB relies on a single collection since all the trips and polygons are stored in a single collection (i.e. nyc2015).

The queries have been executed on one thousand random pickup locations. Since the detected polygons match with each other, another test to assess the spatial accuracy of the results have not been conducted. However, there is more than two fold difference in execution times favouring MongoDB, which is illustrated in Fig. 5.

One of the use-cases of pip query is to manage taxi service in an effective way. For instance, all the pickups that occurred within a given period could be investigated to have a better plan for the upcoming period. The pip analysis conducted on 22

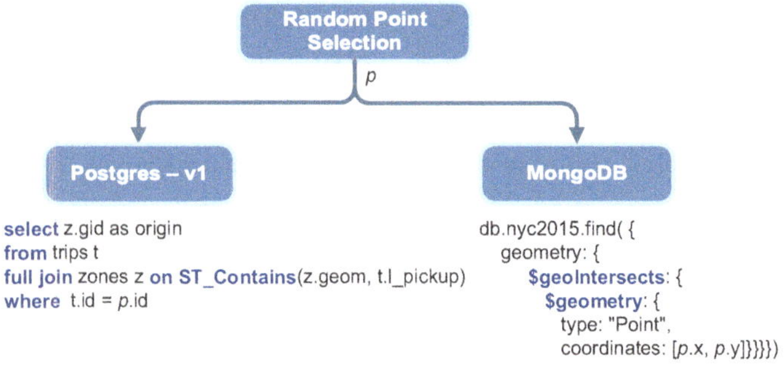

Fig. 4 Formalisation of pip queries in Postgres and MongoDB

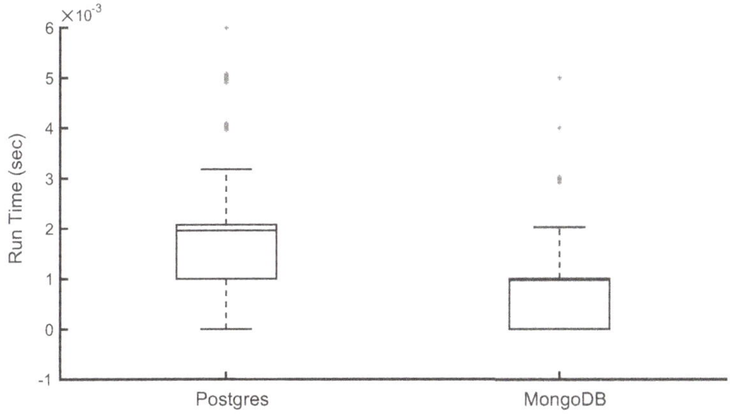

Fig. 5 Run times of the pip query

August 2015 are visualized in Fig. 6 demonstrate the substantial spatial variation in taxi demand.

Amongst the total 357,435 pickups that occurred on 22 August 2015, more than 90% of them occurred in the Manhattan Borough. Two airports, John F. Kennedy

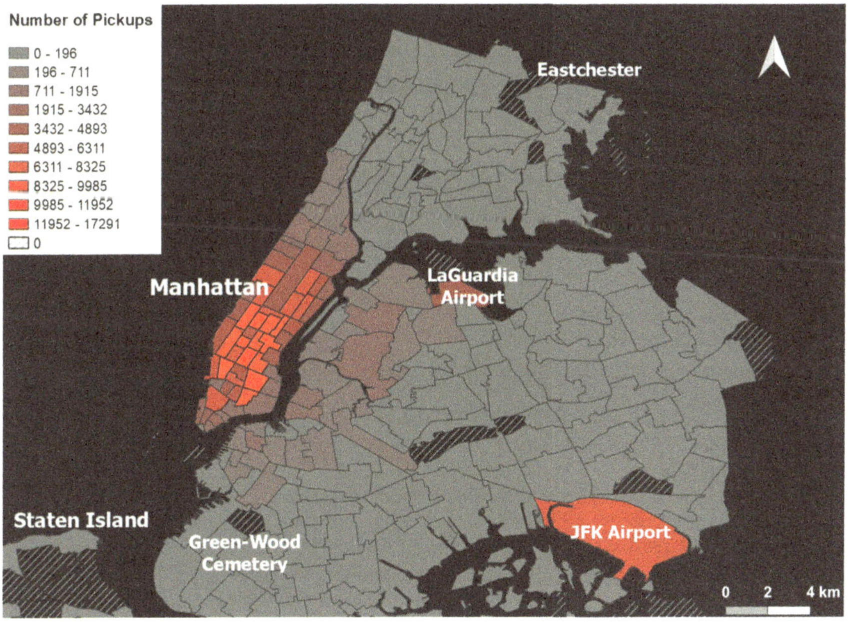

Fig. 6 Thematic map of number of pickups that occurred on 22 August 2015

(JFK) and LaGuardia also attracted considerable demand. No taxi pickups occurred on several taxi zones during the investigated day such as Eastchester and Green-Wood Cemetery.

3.4 Journey Time Analysis

Another use-case of the pip query is journey time analysis between different regions (or zones) of a city. In this way, passengers and companies could plan their journeys accordingly. This section utilises Postgres to investigate the journey time variation between John F. Kennedy (JFK) and LaGuardia (LGA) airports. It is evident that the journey times between these two zones are dependent on the time of day. The rationale behind relying on Postgres for this purpose is the simplicity to encompass the desired analysis within a single query. On the other hand, in MongoDB, two queries must be executed to determine the origin and dropoff zones of a trip, and similarly another query would be required to determine the temporal aspect of the trip (i.e. whether the trip occurred on a weekday or weekend and at which time interval of day the pickup occurred).

A Python script is written by relying on the PostGIS query illustrated in Fig. 7 to investigate the journey time variations between the two aforementioned airports. The script accepts temporal resolution and the analysis interval (i.e. *i_start* and *i_end*) as inputs, and based on them it generates the relevant variables used in the query. For example, if the temporal resolution is set to be 60 min, then the query would be executed 24 times, since daily journey time series is generated. In addition, weekdays and weekends can be analysed separately. The variable *dow* (i.e. day of week) is a set containing values (1, 2, 3, 4, 5) or (0, 6) if the analysis is conducted for weekdays and weekends respectively. Note that, this query is a specialization of the Postgres query described in Fig. 4, in which temporal aspect is incorporated to pip.

The journey times that occurred between JFK and LGA are recorded on an hourly basis for the weekdays and weekends of 2015. There is a substantial variation in journey times. In order to improve legibility, those journeys that took more than 75 min are omitted in Fig. 8. It is also important to highlight that the number of trips in each time interval are different due to the variations in travel demand. This variation

```
select z1.id as O, z2.id as D, (t_dropoff - t_pickup)
from trips
full join zones z1 on st_contains(z1.geom, l_pickup)
full join zones z2 on st_contains(z2.geom, l_dropoff)
where t_pickup >= '2015-01-01' and t_pickup < '2016-01-01'
  and (extract(hour from t_pickup) between hr_start and hr_end)
  and (extract (minute from t_pickup) between min_start and min_end)
  and z1.gid = O and z2.gid = D
  and extract (dow from t_pickup) in dow
```

Fig. 7 PostGIS query to determine the journey times between a given OD pair

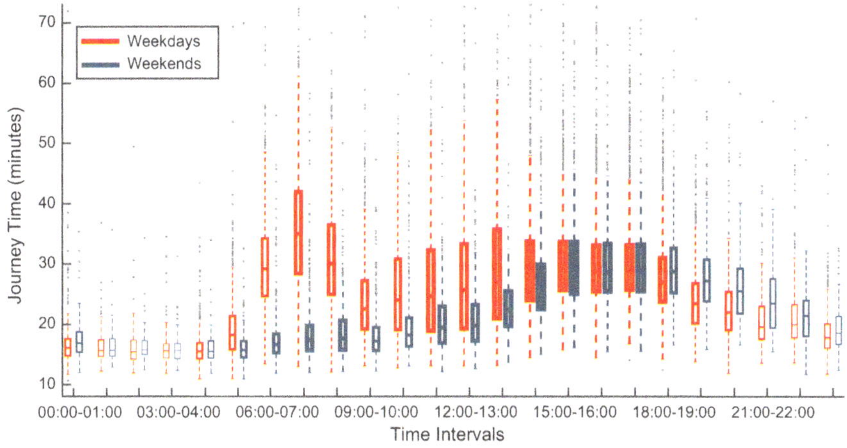

Fig. 8 Journey time distribution of taxi trips for the year 2015 between JFK and LGA

is considered by changing the thickness of the boxplot based on the number of trips that occurred at that interval. The thickness of the boxes are considered separately for weekdays and weekends. Specifically, the lowest number of trips from JFK to LGA is 120 and 54 for weekdays and weekends (both between 02:00 and 03:00) respectively, but they have the same thickness on the plot.

Several observations could be made from this visualisation. First, the difference between a weekday and weekend commute could be substantial, especially in the morning peak period. In addition, on weekdays, morning peak causes higher journey times, whereas on weekends its the evening peak. Overall, the average longest journeys are approximately 35 and 29 min for weekdays (between 07:00 and 08:00) and weekends (between 15:00 and 16:00) respectively. On the other hand, the fastest journeys take approximately 16 min for both weekdays and weekends between 04:00 and 05:00. Second, taxi demand is the highest between 14:00 and 18:00, which would probably match with the flight arrival times. This four-hour interval accounts for approximately one-third of all the taxi trips between JFK and LGA for both weekdays and weekends. Third, though the difference is not substantial, it is possible to have longer journey times during weekends. This difference can be attributed to the increased nightlife activities of the city. At its height, between 21:00 and 22:00, the average journeys take approximately 23.5 min, which is 3.5 min longer than the weekdays.

4 Conclusions

Recent advances in sensor technologies allow researchers and practitioners collect vast amounts of spatial data on a daily basis. In order to analyse and query such large datasets, it is often necessary to use a DBMS. Consequently, most of the GIS courses being thought include a DBMS component. However, traditional way to teach a GIS module is based on relational DBMSs. Specifically, data are stored on tables that are inherently related with each other. This relation could be due to common attributes or location. The emergence of web-based GIS; however, encourages researchers and analysts to investigate NoSQL DBMSs. Consequently, there is a growing need to incorporate the terminology and concepts of NoSQL DBMSs within a GIS module.

The aim of this chapter is to develop an open-source library to analyse the spatial query performance of two commonly used DBMSs, Postgres and MongoDB, on an openly available big origin–destination dataset [5]. Due to their wide use in different contexts ranging from taxi sharing to demand analysis, the performance of KNN and pip queries is investigated. In both of the queries, MongoDB outperformed Postgres in terms of run-time. Regarding the KNN query, regardless the value of k, MongoDB instantly detects the neighbouring points of a query point. On the other hand, run-time increases exponentially in Postgres as k increases. In addition, Postgres misses some of the closer neighbours in KNN query. Therefore, MongoDB is better in terms of spatial accuracy when detecting the KNN of a query point as well. In addition, regarding the pip query, MongoDB is almost two-fold faster than Postgres. These results suggest that spatial index used in MongoDB is more effective than the one realized in Postgres. It should be noted that these findings depend on the investigated versions of Postgres and MongoDB, and subject to change due to the rapid update cycles of these technologies. This finding is important as teaching material on GIS can be based on relational DBMSs [42]. The developed library could be used to facilitate teaching of both relational and NoSQL DBMS through investigating a real-life dataset. In addition, the library could be extended to include queries that are more complex as well as incorporate other DBMSs.

This research could be extended in different directions. First, line objects could be included within the analysis. Utilizing roads, for instance, could be valuable in different aspects including data quality analysis and map matching. Second, response times in a web-based environment could be evaluated. Third, 'horizontal scalability' of MongoDB could be investigated for further performance gains. All of these possible research directions could be integrated into term projects of students. The open-source nature of the analysed data as well as the developed library could facilitate further development in this regard.

Acknowledgements This research is supported by the Scientific and Technological Research Council of Turkey (TÜBİTAK) with the grant number of 118Y282. The developed library is tested by Buğra Coşkun. Initial contributors are Çağlayan Aşkın and Serdar Darendelioğlu.

References

1. Ma, X., Wang, Y.: Development of a data-driven platform for transit performance measures using smart card and GPS data. J. Transp. Eng. **140**, 04014063 (2014). https://doi.org/10.1061/(ASCE)TE.1943-5436.0000714

2. Egenhofer, M.J.: Why not SQL! Int. J. Geogr. Inf. Syst. **6**, 71–85 (1992). https://doi.org/10.1080/02693799208901897
3. Tudorica, B.G., Bucur, C.: A comparison between several NoSQL databases with comments and notes. In: 2011 RoEduNet International Conference 10th Edition: Networking in Education and Research, pp. 1–5 (2011). https://doi.org/10.1109/RoEduNet.2011.5993686
4. TLC: NYC Taxi & Limousine Commission—Trip Record Data. https://www.nyc.gov/html/tlc/html/about/trip_record_data.shtml. Accessed 2018/08/04
5. Anbaroglu, B.: Query-Performance-Analyses. https://github.com/banbar/Query-Performance-Analyses. Accessed 2019/08/12
6. Wan, L., Huang, Z., Peng, X.: An effective NoSQL-based vector map tile management approach. ISPRS Int. J. Geo-Inf. **5**, 215 (2016). https://doi.org/10.3390/ijgi5110215
7. Kumar, K.B.S., Srividya, Mohanavalli, S.: A performance comparison of document oriented NoSQL databases. In: 2017 International Conference on Computer, Communication and Signal Processing (ICCCSP), pp. 1–6 (2017). https://doi.org/10.1109/ICCCSP.2017.7944071
8. Kim, Y.S., Kim, T., Carey, M.J., Li, C.: A comparative study of log-structured merge-tree-based spatial indexes for big data. In: 2017 IEEE 33rd International Conference on Data Engineering (ICDE), pp. 147–150 (2017). https://doi.org/10.1109/ICDE.2017.61
9. Gupta, N., Agrawal, R.: NoSQL security. Adv. Comput. (2018). https://doi.org/10.1016/bs.adcom.2018.01.003
10. Lübcke, A., Schäler, M., Köppen, V., Saake, G.: Relational on demand data management for IT-services. In: 2014 IEEE Eighth International Conference on Research Challenges in Information Science (RCIS), pp. 1–12 (2014). https://doi.org/10.1109/RCIS.2014.6861078
11. Cho, W., Choi, E.: A basis of spatial big data analysis with map-matching system. Cluster Comput. **20**, 2177–2192 (2017). https://doi.org/10.1007/s10586-017-1014-1
12. Matuszka, T., Kiss, A.: Experimental evaluation of some geodata management systems. In: 2014 9th International Conference on Computer Engineering Systems (ICCES), pp. 92–97 (2014). https://doi.org/10.1109/ICCES.2014.7030935
13. Schmid, S., Galicz, E., Reinhardt, W.: WMS performance of selected SQL and NoSQL databases. In: International Conference on Military Technologies (ICMT) 2015, pp. 1–6 (2015). https://doi.org/10.1109/MILTECHS.2015.7153736
14. INSPIRE: Technical Guidance for the Implementation of INSPIRE Download Services Using Web Coverage Services (WCS) | INSPIRE, p. 53. https://inspire.ec.europa.eu/id/document/tg/download-wcs. Accessed 2019/11/02
15. Agarwal, S., Rajan, K.S.: Performance analysis of MongoDB versus PostGIS/PostGreSQL databases for line intersection and point containment spatial queries. Spat. Inf. Res. **24**, 671–677 (2016). https://doi.org/10.1007/s41324-016-0059-1
16. Coşkun, İ.B., Sertok, S., Anbaroğlu, B.: K-nearest neighbour query performance analyses on a large scale taxi dataset: PostgreSQL vs. MongoDB. ISPRS Int. Arch. Photogramm. Remote Sens. Spat. Inf. Sci. 1531–1538 (2019). https://doi.org/10.5194/isprs-archives-XLII-2-W13-1531-2019
17. Bakillah, M., Lauer, J., Liang, S.H.L., Zipf, A., Arsanjani, J.J., Mobasheri, A.: Exploiting big VGI to improve routing and navigation services. In: Big Data: Techniques and Technologies in Geoinformatics. Taylor & Francis (2014). https://doi.org/10.1201/b16524-12
18. Li, S., Dragicevic, S., Castro, F.A., Sester, M., Winter, S., Coltekin, A., Pettit, C., Jiang, B., Haworth, J., Stein, A., Cheng, T.: Geospatial big data handling theory and methods: a review and research challenges. ISPRS J. Photogramm. Remote Sens. **115**, 119–133 (2016). https://doi.org/10.1016/j.isprsjprs.2015.10.012
19. Brovelli, M.A., Kilsedar, C.E., Zamboni, G.: Visualization of VGI data through the new NASA web world wind virtual globe. Int. Arch. Photogramm. Remote Sens. Spat. Inf. Sci. **XLI-B4**, 205–209 (2016). https://doi.org/10.5194/isprsarchives-XLI-B4-205-2016
20. Mobasheri, A., Sun, Y., Loos, L., Ali, A.L.: Are crowdsourced datasets suitable for specialized routing services? Case study of OpenStreetMap for routing of people with limited mobility. Sustainability **9**, 997 (2017). https://doi.org/10.3390/su9060997

21. Sánchez-Cambronero, S., Jiménez, P., Rivas, A., Gallego, I.: Plate scanning tools to obtain travel times in traffic networks. J. Intell. Transp. Syst. **21**, 390–408 (2017). https://doi.org/10.1080/15472450.2017.1298037
22. Barann, B., Beverungen, D., Müller, O.: An open-data approach for quantifying the potential of taxi ridesharing. Decis. Support Syst. **99**, 86–95 (2017). https://doi.org/10.1016/j.dss.2017.05.008
23. Zhu, Y., Ozbay, K., Xie, K., Yang, H.: Using big data to study resilience of taxi and subway trips for Hurricanes Sandy and Irene. Transp. Res. Rec. J. Transp. Res. Board **2599**, 70–80 (2016). https://doi.org/10.3141/2599-09
24. Peng, S., Samet, H.: Analytical queries on road networks: an experimental evaluation of two system architectures. In: Proceedings of the 23rd SIGSPATIAL International Conference on Advances in Geographic Information Systems, pp. 1:1–1:10. ACM, New York, NY, USA (2015). https://doi.org/10.1145/2820783.2820806
25. Hrnčíř, J., Rovatsos, M., Jakob, M.: Ridesharing on timetabled transport services: a multiagent planning approach. J. Intell. Transp. Syst. **19**, 89–105 (2015). https://doi.org/10.1080/15472450.2014.941759
26. Tang, J., Liu, F., Wang, Y., Wang, H.: Uncovering urban human mobility from large scale taxi GPS data. Physica A **438**, 140–153 (2015). https://doi.org/10.1016/j.physa.2015.06.032
27. Zhang, X., Deng, Y., Li, Q., Skitmore, M., Zhou, Z.: An incident database for improving metro safety: the case of Shanghai. Saf. Sci. **84**, 88–96 (2016). https://doi.org/10.1016/j.ssci.2015.11.023
28. Zhuang, L., Gong, J., He, Z., Xu, F.: Framework of experienced route planning based on taxis' GPS data. In: 2012 15th International IEEE Conference on Intelligent Transportation Systems, pp. 1026–1031 (2012). https://doi.org/10.1109/ITSC.2012.6338867
29. Li, J.Q., Zhou, K., Zhang, L., Zhang, W.B.: A multimodal trip planning system with real-time traffic and transit information. J. Intell. Transp. Syst. **16**, 60–69 (2012). https://doi.org/10.1080/15472450.2012.671708
30. Nguyen, T.T.: Indexing PostGIS databases and spatial query performance evaluations. Int. J. Geoinform. **5**, 1–9 (2009)
31. Gebaly, K.E., Aboulnaga, A.: Robustness in automatic physical database design. In: Proceedings of the 11th International Conference on Extending Database Technology: Advances in Database Technology, pp. 145–156. ACM, New York, NY, USA (2008). https://doi.org/10.1145/1353343.1353365
32. Biau, G., Devroye, L.: Lectures on the Nearest Neighbor Method. Springer International Publishing (2015)
33. Cracknell, M.J., Reading, A.M.: Geological mapping using remote sensing data: a comparison of five machine learning algorithms, their response to variations in the spatial distribution of training data and the use of explicit spatial information. Comput. Geosci. **63**, 22–33 (2014). https://doi.org/10.1016/j.cageo.2013.10.008
34. Cai, P., Wang, Y., Lu, G., Chen, P., Ding, C., Sun, J.: A spatiotemporal correlative k-nearest neighbor model for short-term traffic multistep forecasting. Transp. Res. Part C Emerg. Technol. **62**, 21–34 (2016). https://doi.org/10.1016/j.trc.2015.11.002
35. Zhong, R., Li, G., Tan, K., Zhou, L., Gong, Z.: G-tree: an efficient and scalable index for spatial search on road networks. IEEE Trans. Knowl. Data Eng. **27**, 2175–2189 (2015). https://doi.org/10.1109/TKDE.2015.2399306
36. Hormann, K., Agathos, A.: The point in polygon problem for arbitrary polygons. Comput. Geom. **20**, 131–144 (2001). https://doi.org/10.1016/S0925-7721(01)00012-8
37. Kumar, G.N., Bangi, M.: An extension to winding number and point-in-polygon algorithm. IFAC-PapersOnLine **51**, 548–553 (2018). https://doi.org/10.1016/j.ifacol.2018.05.092
38. Zhang, J., You, S.: Speeding up large-scale point-in-polygon test based spatial join on GPUs. In: Proceedings of the 1st ACM SIGSPATIAL International Workshop on Analytics for Big Geospatial Data, pp. 23–32. ACM, New York, NY, USA (2012). https://doi.org/10.1145/2447481.2447485

39. Naziha, A., Fu, L., Mohamed Elamine, G., Wang, L.: A method to construct an indoor air pollution monitoring system based on a wireless sensor network. Sensors **19**, 967 (2019). https://doi.org/10.3390/s19040967
40. Zhang, J., You, S., Gruenwald, L.: Efficient parallel zonal statistics on large-scale global biodiversity data on GPUs. In: Proceedings of the 4th International ACM SIGSPATIAL Workshop on Analytics for Big Geospatial Data, pp. 35–44. ACM, New York, NY, USA (2015). https://doi.org/10.1145/2835185.2835187
41. Bakker, T., van Asselt, K., Bontsema, J., Müller, J., van Straten, G.: Autonomous navigation using a robot platform in a sugar beet field. Biosyst. Eng. **109**, 357–368 (2011). https://doi.org/10.1016/j.biosystemseng.2011.05.001
42. King, M.A., Arnette, A.N.: Integrating geographic information systems in business school curriculum: an initial example. Decis. Sci. J. Innov. Educ. **9**, 325–347 (2011). https://doi.org/10.1111/j.1540-4609.2011.00318.x

Investigating the Use of Historical Node Location Data as a Source to Improve OpenStreetMap Position Quality

Talia Dror, Yerach Doytsher, and Sagi Dalyot

Abstract Past years have witnessed the rapid use of participatory mapping and the establishment of open-source mapping infrastructures. The collection of user-generated geographic content has proven to provide highly effective means of acquiring useful and detailed global environmental data. OpenStreetMap is perhaps the best-established and reliable example that makes use of crowdsourced maps, constantly and rapidly evolving. Since data is heterogeneous in nature, the position quality of the objects edited and mapped into OpenStreetMap varies. The current OpenStreetMap practice is that the last node location edit of a map-feature is presented in the map, possibly inferring that it is the most accurate representation. Accordingly, OpenStreetMap presents the recent version of the accumulated edited and mapped data, disregarding the earlier versions. This chapter investigates the positional quality of historical node versions, evaluating the current OpenStreetMap practice alongside alternative spatiotemporal data models for improved node location calculation. With the proper incremental use of the historical node versions, the more accurate location for close to 60% of all the analyzed nodes was automatically selected between the last version location and the location calculated by the best model. This statistic increases to 76% for nodes that show larger discrepancies when compared to the reference locations. These preliminary results challenge the current OpenStreetMap practice by suggesting an alternative data model that use historical versions. The outcome validates and extends the existing patchwork approach associated with the crowdsourcing of user-generated geographic data and information that are the basis of OpenStreetMap.

Keywords Location accuracy · OpenStreetMap (OSM) · Spatial statistical analyses · Historical geodata

---

T. Dror · Y. Doytsher · S. Dalyot (✉)
Mapping and Geo-Information Engineering, 32000 Technion, Haifa, Israel
e-mail: dalyot@technion.ac.il

A. Mobasheri (ed.), *Open Source Geospatial Science for Urban Studies*, Lecture Notes in Intelligent Transportation and Infrastructure,
https://doi.org/10.1007/978-3-030-58232-6_4

1 Introduction

Crowdsourcing is one of the most significant and influential paradigms existing today, increasingly expanding to many new scientific disciplines, mainly thanks to the revolution of Web 2.0 (e.g., [1–3]). In the mapping and geo-information disciplines, crowdsourcing refers to a large group of people that can substitute surveying experts and professional cartographers in their mapping tasks by contributing user-generated geographic content, e.g., location data and information, replacing the existing top-down mapping system with a bottom-up one [4]. This results in the increasing numbers of location-based projects, applications and services that make use of the contributed crowdsourced user-generated data and information.[1]

One of the most promising Volunteered Geographic Information (VGI) projects is OpenStreetMap (OSM), in which contributing groups and volunteers update the geospatial database of this online open-source map with new and updated geographic content. OSM contains vector data created, updated and maintained by the public: as of June 2019, the number of registered OSM members is more than 5.45 million.[2] The geographic and attributed data is easily edited and downloaded, whereas open-source codes are provided to process and analyze the existing mapping data. The OSM project is spreading widely and rapidly, easily accessed by individuals and small organizations, especially in countries where geographic information is expensive or sparse, as well as in areas where the mapping infrastructure is evident to update frequently [5, 6].

OSM contains node and linear elements featuring the most current state of the contributed mapping data [7]. OSM introduces an event-based map update, as opposed to a cyclic one, meaning that if a location or an attribute of an existing map element is surveyed and mapped to the OSM database, it replaces the data that exists in the displayed map. OSM database contains and archives all data and events from its initial establishment in 2004. The complete OSM history data is termed "Full History Planet File",[3] and is updated continuously by volunteers according to their immediate knowledge of their surroundings. Nonetheless, OSM merely displays the most recent version of the accumulated data [8, 9]. All other versions are stored in the "Full History Planet File", not visible to the map users. For example, as of June 2019, approximately 5.2 Billion nodes[4] are presented in OSM out of close to 6.5 Billion nodes that are the complete versions existing in the full database. The current rate of new OSM nodes that are created according to the existing statistics is approximately 2 million per day, whereas the number of the modified nodes, i.e., the updated information for nodes that exist in the OSM database, reaches 500 thousand per day (see Footnote 4).

Current research mostly investigates OSM's positional accuracy and completeness, e.g., nodes that represent points of interest (POI) and lines that represent roads

[1] wiki.openstreetmap.org/wiki/List_of_OSM-based_services.

[2] https://osmstats.neis-one.org

[3] https://planet.osm.org/planet/full-history/

[4] https://osmstats.neis-one.org/?item=nodes

and buildings of all types. This means that OSM data is analyzed in respect to a specific time-stamp, practically the last version is examined and compared to an accurate authoritative mapping data reference. Although researchers prove that the location accuracy and completeness measures of OSM are good and continue to improve, not many have addressed these quality measures on a temporal level, i.e., historical versions. Meaning, how does OSM data improves over time, by examining the temporal positional change of the edited geographic entities.

Another important aspect is that OSM displays the last edited data version only, considered as the most relevant and accurate, without taking into consideration the existing historical version data. The assumption is that the 'surveying devices' (mostly relying on mobile consumer GNSS sensors) and mapping infrastructures (mostly satellite-based orthoimages base-maps) continue to improve, thus a new edit should provide more accurate observations. This premise does not always stand, since user-generated data is heterogonous by nature, affecting the position quality of the objects edited and mapped. Such that a question arises as to how qualitative this practice is in terms of the positional accuracy, and whether it can be handled differently by using all—or part of—the accumulated historical version data. This chapter examines and evaluates the use of three data models that rely on mean, weighted mean, and cluster-based calculations of the historical location data of point objects. The results show that the current practice of using the last version does not always stand, and implementing an alternative practice that uses the historical data has the potential of improving OSM position accuracy.

2 Literature Overview

According to Senaratne et al. [10], 17 quality measures can be analyzed and assessed in map credibility that is based on VGI; still, the majority of research that is related to OSM focus on the positional accuracy measure. Already in 2010, Haklay [3] analyzed OSM accuracy and completeness of linear objects representing motorways in relation to an authoritative map. This was based on a geometrical-based buffering, showing that OSM data had an average positional discrepancy value of 6 m. Since then, other studies that examined these aspects in relation to linear objects for different areas, while implementing various quality analyses methodologies, came to similar conclusions—showing that for urban areas both measures continue to improve, whereas for rural areas and developing regions the measures showed inferior results (e.g., [11–19]).

Research examining the node positional accuracy of the OSM features, such as POIs (e.g., hospitals, schools, etc.) and road junctions, used the Euclidian distance measure from reference point objects, e.g., [20, 21]; the latter found similar average discrepancy values of positional accuracy of fewer than 7 m. Al-Bakri and Fairbairn [22] compared the point data of OSM and a reference 'ground truth' field survey dataset using the RMSE measure to estimate OSM's positional accuracy. The results showed that the positional accuracy has large discrepancy values, i.e., geometrically they do not match the reference dataset used, with values larger than 10 m. Helbich et al. [23] presented a spatial statistical comparative method to compute the positional accuracy of the OSM road junctions by comparing them to precise field survey data and commercial Tele Atlas dataset. Analyzing a well-mapped city in Germany, the results showed that both the OSM and the Tele Atlas datasets have similar spatial deviations from the field survey data. Furthermore, the OSM data showed spatial heterogeneity in the positional error distribution, leading to significant clusters of high and low positional accuracy. As in Jackson et al. [24], the authors concluded that according to their results, the OSM data can be used for small and medium scale mapping applications.

The aforementioned studies focused on examining the OSM data accuracy at a specific timestamp, all analyzing the most current and updated available version of OSM database. The authors relied on the assumption that the OSM datasets used for analyses are certain to show the most accurate and reliable existing mapping data when compared to an external authoritative sources. Only rarely the examination and investigation of historical OSM data are addressed. One example is Haklay et al. [25], analyzing whether the positional accuracy of intersection features improve as the number of contributors increases. The authors have concluded that the overall aggregated location accuracy did improve, although no effort was made to study whether previous versions can contribute to the overall positional accuracy. Mashhadi et al. [26] have analyzed the OSM POIs location accuracy with respect to the number of edits each had. The authors concluded that the higher the number of POI edits has, the higher its quality is, while the quality of the contributions in OSM is independent of the number of edits they have undergone. Mooney and Corcoran [8] examined OSM ways (linear features) having 15—and more—edits in respect of tagging, i.e., validating attribute and semantic completeness of the features. The authors showed that only a small number of contributors created or edited most of the existing data, without finding a strong correlation between the number of versions and the completeness value. Corcoran and Mooney [27] analyzed the OSM street network edited during the years 2007–2011. The authors focused on metric and topological characteristics, finding that only some characteristics were similar in each region.

Vandecasteele and Devillers [28] explored the use of semantic similarity of historic features, introducing a tag recommendation system (OSMantic) that automatically suggests relevant and appropriate tags to contributors. This was made to improve the semantic quality and reduce the semantic heterogeneity of the OSM tagging. Keßler and Groot [29] introduced a set of trust measures, which are derived from the feature provenance, such as the number of versions, contributors, confirmations, etc. To evaluate the trust measures, the results were compared to the quality measures obtained from a field survey, proving that the user-generated feature quality can be assessed by using a trust model that is based on the data provenance. The quality measures were based on the thematic accuracy, topological consistency, and information completeness, with no positional accuracy assessment. Zhao et al. [30] introduced a spatiotemporal VGI model that considers the contributor reputation and object trustworthiness, among others: the contributor's experience, contributor's trust and version. Although the authors did not asses the overall location accuracy improvement, they concluded that keeping the last feature created by a more trustworthy contributor is not always better, and that the computation of the object trustworthiness and user reputation is still an open issue. This conclusion is also demonstrated in Nasiri et al. [31], showing that using historical data of ways (road centerlines) with Voronoi diagram to filter data outliers improved the overall completeness and location quality of these OSM features. To sum, a more comprehensive analysis of the OSM historical location data is required, assessing its validity on the overall OSM position accuracy.

3 Methodology

Three data models are implemented that rely on an aggregative process and classification methods that make use of the historical data of node versions. The models are evaluated according to the number of updates (versions), i.e., validating whether the node locations are improved with respect to the reference data, and what is the most qualitative and probable solution to achieve the most accurate overall node location model in OSM.

3.1 Mean Calculation

This model uses a simple location calculation that is based on the general mean function applied on all coordinate versions of a specific node (see Eq. 1 with $w_i = 1$, which denotes weight). No preliminary considerations or constraints are made, assuming all versions have equal weight and serve as a valid data chunk in the process.

3.2 Weighted Mean Calculation

In this model, we assume there exists a correlation between the version number and the node location quality. Meaning that latter updates are more accurate than earlier ones due to the mapping infrastructure used (e.g., orthoimages) and/or the used data collection device (GNSS sensor) that continue to advance. However, no metadata documentation regarding the user's mapping method and practice of a certain entity is stored in the OSM repository e.g., GNSS trajectory, satellite imagery, etc. [5, 32]. Accordingly, earlier versions (i in Eq. 1) receive a lesser weight value when compared to more recent versions that receive a higher weight value. Meaning that weight is applied to all node location versions, and is derived from the version index i in relation to all versions exist for an object, i.e., the higher version value—the higher weight (significance) it receives in the model.

$$X_{WM} = \frac{\sum_{i=1}^{n} w_i \cdot X_{\text{ver}_i}}{\sum_{i=1}^{n} w_i} \quad (1)$$

where X_{WM} is the X ordinate of a node after the weighted mean calculation, X_{ver_i} is the ordinate of the ith version, w_i is the weight of the ith version, where $i = 1$ to n (the number of node versions). The same is carried out for the Y ordinate.

3.3 Cluster-Based Calculation

This model calculates a majority voting measure that is based on clusters of similar locations of different versions, and is based on the Density-Based Spatial Clustering of Applications with Noise (DBSCAN) algorithm [33]. This algorithm was chosen since it is a density-based one that does not require a predefined parameter for the number of predicted clusters, but only a minimum number of nodes in a cluster (*Nop*), and a radius for neighborhood points (*Eps*). The *Nop* value is defined as a minimum of three points for a cluster. The *Eps* value is varying from node to node derived from the density of the existing node locations, relying on the notion of Varied DBSCAN [34]. *Eps* is analyzed separately for each node, using the Euclidian distance matrix from each version to all others, choosing the minimum value from the median values of every row.

The resulting clusters are analyzed with respect to spatial distribution thresholds—using the cluster area and perimeter values. For example, a wide distribution of versions is assumed to resemble a case having no location similarities, assuming that the last version should be chosen. Alternatively, clusters having dense distribution prove a strong correlation between the different node location versions. The threshold values used are: (1) *area* is below double the average area value of all the convex hulls constructed from the location version list of all nodes; and, (2) *shapeGeom*, depicted in Eq. 2, is below 1/3. The latter threshold resembles a measure of shape compactness that can help in identifying a non-complex spatially homogenous cluster. Value of 1.0 denotes a compact shape (circle), thus a value of 1/3 was chosen to give a degree of freedom to the cluster shape complexity. For very dense clusters, e.g., cluster area less than 1 m^2, the above-mentioned values are not used.

$$\text{shapeGeom} = \frac{4 \cdot \pi \cdot a}{P^2} \quad (2)$$

where a is the cluster area, and P is the cluster perimeter.

This model enables the filtering of locations (versions) that are considered as outliers with respect to other locations, which were clustered, thus improving the overall value of the estimated node location. For clustered node versions, the weighted mean model in Eq. 1 is used to calculate the node coordinates.

In case of two node versions, the weighted mean model is used as long as the distance between them is less than 20 m. In case none of the above stands, the last

Fig. 1 Pseudo code for cluster-based model

```
Cluster-based model:
--------------------------------------------------------------
Input: A list of nodes list pntListOfLists
Output: The result node coordinates resultPnt

for each pntList in pntListOfLists
    if (pntList.Size > 2) \\ in case of more than 2 versions
            clusteredList ← DBSCAN clustering
            resultPnt ← weighted mean of clusteredList
            area ← calculation of area for clusteredList
            perimeter ← calculation of perimeter for clusteredList

            if ((area > 1) or (perimeter > avgPerimiter/2))
                if ((area > avgArea*2) or (shapeGeom < 1/3))
                        resultPnt ← last version node
                end
            end

    else \\ in case of 2 versions
            length ← calculation of the Euclidian distance of pntList
            if (length > 20)
                    resultPnt ← last version node
            else
                    resultPnt ← weighted mean of pntList
            end
    end
return resultPnt
```

version is used as the computed node location. The pseudo code of this approach is depicted in Fig. 1.

Figure 2 depicts two examples of the cluster-based model. The top image (a) shows a cluster (red triangle) having an *area* value that is smaller than the threshold, but with a *shapeGeom* value that is bigger than the threshold, thus the last version (red circle) is chosen. The bottom image (b) shows a cluster having both the aforementioned thresholds met, thus the weighted average location (red circle) is chosen as the most probable location. For both examples, the model automatically computes node locations that are more accurate (closer) in respect to the reference location (green asterisk)—the last version (a) and the mean average (b), respectively.

4 Experimental Results

4.1 Data Preparation

The evaluation is conducted for an area in The Technion, Haifa. Retrieving the historical archives of the OSM objects, i.e., location data with edits (versions), was made using a modified version of the process described in Mooney and Corcoran [8, 35]. The OSM data "Latest Full History Planet PBF File" was downloaded[5] containing

[5]https://planet.osm.org/planet/full-history/

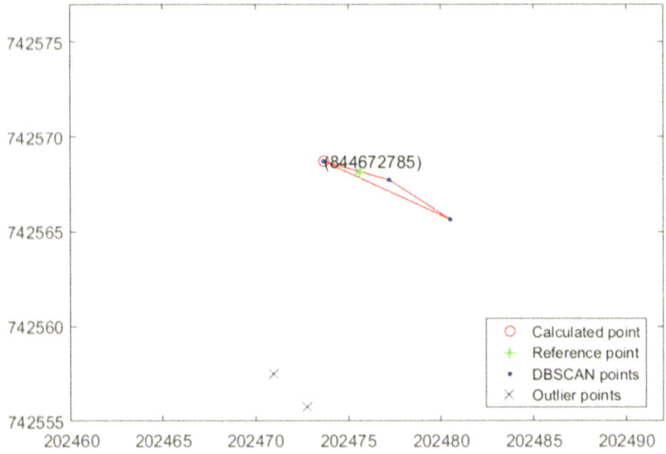

(a)

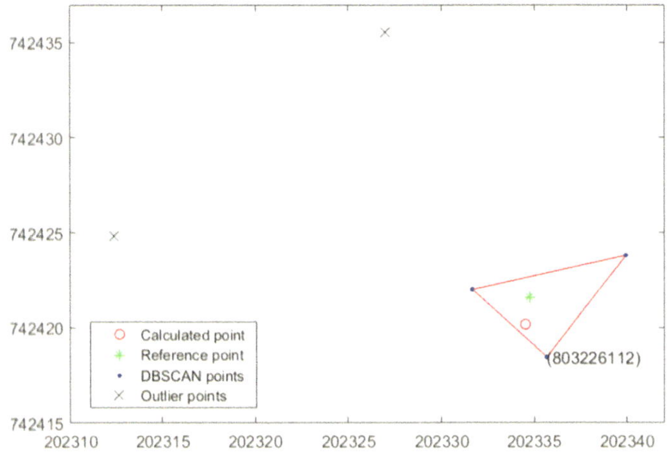

(b)

Fig. 2 Two examples of the cluster-based model: **a** thresholds are not met thus the last version is chosen (the last version is denoted with the id number); **b** both thresholds are met, thus the weighted average is chosen. For both, the accurate locations are calculated when compared to the reference location

76 GB of compressed data (1.3 TB uncompressed). For extracting the study area with all the existing edits, the open source osm-history-splitter tool[6] was used on the Ubuntu operating system. The output file was saved in .osm file format, easily uploaded to ArcGIS (using ArcGIS Editor for OpenStreetMap[7]) as 3 shapefiles: points, lines and polygons. A 7-parameter transformation was used to convert the

[6]https://github.com/MaZderMind/osm-history-splitter

[7]https://www.arcgis.com/home/item.html?id=75716d933f1c40a784243198e0dc11a1.

Table 1 Extract of the ArcGIS attribute table displaying the OSM historical data arrangement

OBJECTID	…	OSMID	OSMuser	OSMversion
491		392271182	talkat	1
492		392271182	Momentum	2
493		392271182	Zvika	3
:		:	:	:
1159		787313524	Momentum	1
1160		787313524	Zvika	2
1161		787313524	amitw	3
1162		787313524	masa0021	4
1163		787313524	talkat	5
1164		787313524	Precise Cartography	6

OSM data from geographic coordinates (WGS84) to the local Cartesian coordinate system of Israel Transverse Mercator (ITM).

The data was stored as ArcGIS data table in one list without distinguishing between the various versions of a specific entity, i.e., every row represents a new point entity; the pairing of the "OSMID" and "OSMversion" columns (attributes) represents the desired historical data structure, depicted in Table 1.

After sorting the OSM data—ID (field OSMID) and versions (field OSMversion), reference data is uploaded to ArcGIS, and manual matching is carried out to find corresponding homologous entities. The reference data used is photogrammetric mapping designed for 1:1250 scale, having a positional accuracy of 0.4 m (confidence level of 90%). Nodes representing the building corners (polygon features breakpoints) and the road skeleton vertices are chosen for the analysis. The building corners and road centerline vertices are used since they are relatively easy to identify and match with the reference data. Moreover, since we are familiarized with the analyzed area, and have its historical plans since 2004 (the year of OSM establishment), by choosing these entities we overcome the problem of location changes, as we know that the physical location of the objects did not change due to construction or demolition since they were first mapped. Accordingly, we can assume with very high certainty that the matched nodes were updated by users with the purpose of improving their positional accuracy. Figure 3 depicts two examples of building features from OSM and reference data. Although it is clear that some building node correspondences are easy to identify (yellow nodes), other nodes (dark blue) are more challenging to match since the geometries of the buildings in both databases differ; thus, only homologous nodes are matched for further analyses. Table 2 depicts the 506 matched nodes according to their number of versions, where it is clear that the number decreases as the version number increases.

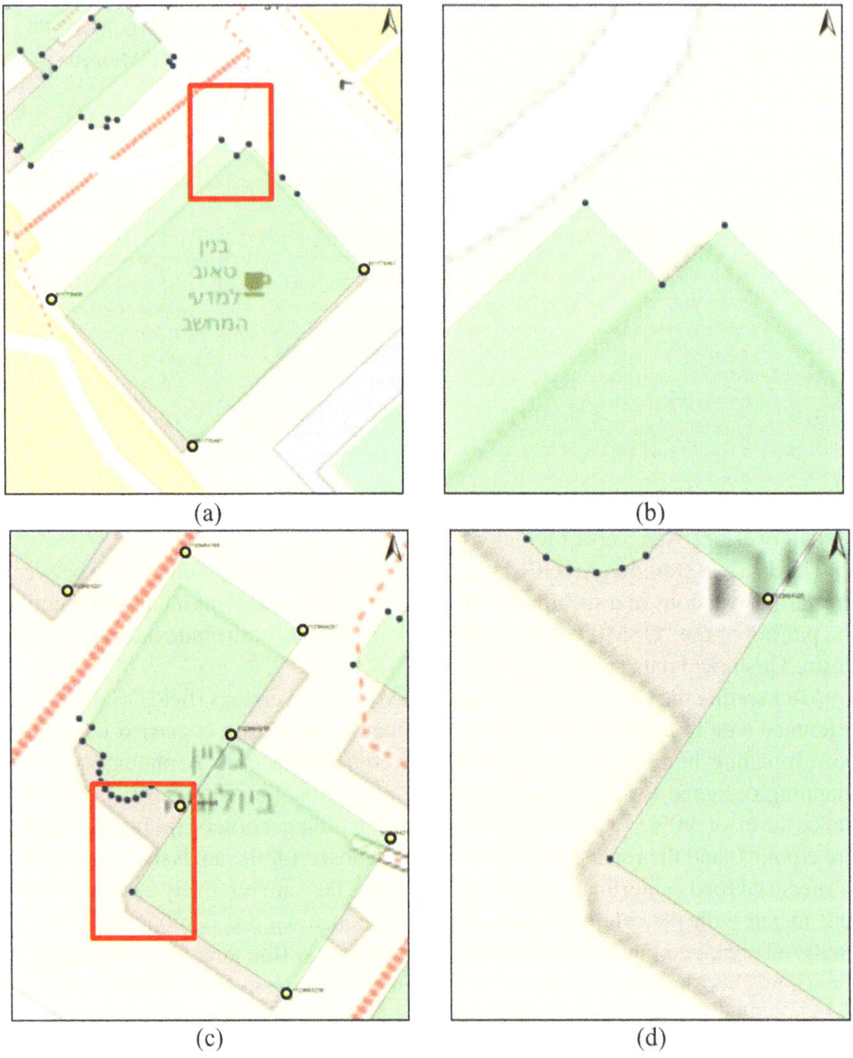

Fig. 3 A comparison of the reference and OSM buildings data. Figures **a** and **c** (in scale of approx. 1:500): reference building footprint (green polygon) and OSM building footprint according to the currently available version (grey polygon). Figures **b** and **d** (in scale of approx. 1:150): zoomed area showing ambiguous matching of building corners (nodes). *Source* www.openstreetmap.org

4.2 Node Location Comparison

Euclidean distance. For the building corners, the Euclidean distance between the OSM nodes (ID and its historical versions) and the matched reference node was used to evaluate the positional discrepancies.

Table 2 The statistics of the number of OSM node elements in the study area of The Technion, Haifa, and the number of homologous OSM and reference data nodes representing building corners and road skeleton vertices

Version	Number of OSM nodes	% from all OSM nodes	Number of matched OSM building/road nodes
1	3529	62.5	–
2	1195	21.2	194
3	477	8.5	149
4	210	3.7	85
5	102	1.8	36
6	69	1.2	24
7 and more	62	1.1	18

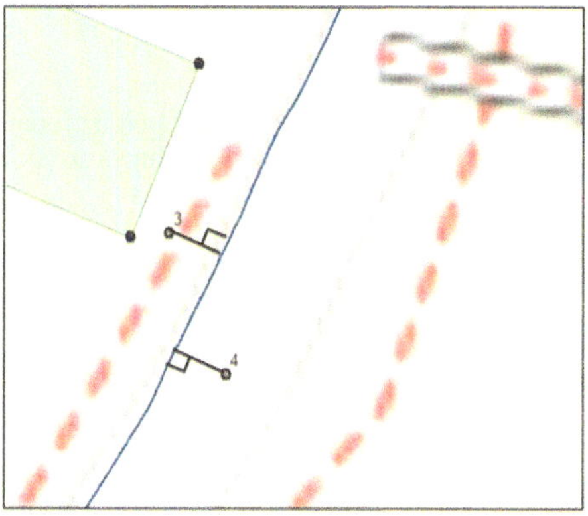

Fig. 4 Point to a line distance. The numbers represent the OSM node version (grey point)

Point to a line distance. For the road centerlines, instead of using the commonly used buffer for calculating the overlap percentage, we used the nearest reference line segment for each OSM node version via the minimum perpendicular distance to evaluate the positional discrepancies, as depicted in Fig. 4.

4.3 Central Limit Theorem

One concern when carrying out this type of statistical analysis is to validate that the sample size serves as a good representation; in this case, the number of matched nodes and the OSM point entity population. For this, we analyzed the population according to the Central Limit Theorem [36], to evaluate whether the data is averaged from 2 arbitrary values: (1) the average of averages, since we use 506 matches having 2–12 versions each, and, (2) randomly chosen n nodes for statistic calculations, where n: $2 \rightarrow 506$. For the theorem to uphold, the average calculated value of a sample of n

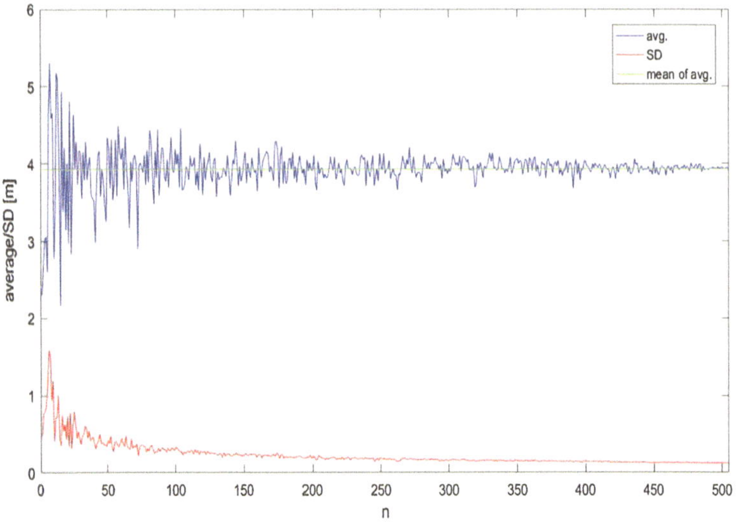

Fig. 5 Validation of the Central Limit Theorem: average (blue) and SD (red) values (Y-axis) calculated for random nodes of n points (X-axis). The mean of averages (green) value converges to the population value

= 30 should converge to a similar value of the entire population, while the SD value should attain a stable value. In this case, the average and the SD values are calculated based on the Euclidean and point to a line distance.

Figure 5 depicts the results, showing that up to $n = 30$ there exist relatively large fluctuations of the average values (blue line), then reduced and converging to a fixed average value of 3.95 m (a similar value was calculated for the sample that includes 506 nodes), and thus equals to the complete population value. Moreover, the SD value also converges to a fixed value of less than 1 m after about 100 points. Accordingly, since both measures converge to a defined value, we ascertain that the Central Limit Theorem holds, and this dataset (sample) can be used for a statistical evaluation.

4.4 Location Accuracy Analysis

A comparison of all node version locations to the corresponding reference nodes was made. Examining the statistics for the Euclidian distances and point to a line distances—buildings vertices and roads centerlines, respectively—which are depicted in Fig. 6, we found that all values show a clear trend of reduction as the number of versions increases (the optimal values are 0, meaning that the OSM nodes correspond completely with the reference nodes). This theoretically verifies the premise of using the last version as the more accurate location. Still, some inconsistencies exist, e.g., from version 3 to 4, and from version 5 to 6, that the general trend

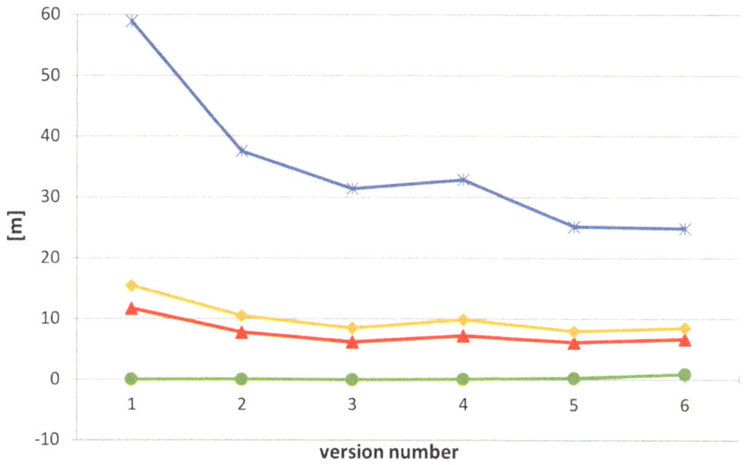

Fig. 6 Statistics for positional accuracy versus version for Euclidian/point to a line distance: mean difference (red), RMSE difference (yellow), the maximum difference (blue), and minimum difference (green)

changes its direction, and values increase. These results coincide with the conclusions made in Haklay et al. [25] regarding the number of contributors with respect to positional accuracy improvement. Although our research examines the number of versions for each node, and not the number of contributors, it can be clearly stated that the number of versions is equal to—or more than—the number of contributors.

Figure 7 depicts an analysis that checks whether the last node version is indeed the accurate location among all other versions in respect to the reference. A clear trend exists that shows that as the number of node versions increases, the percentage

Fig. 7 Percentage values where the last version is the most accurate among other node versions

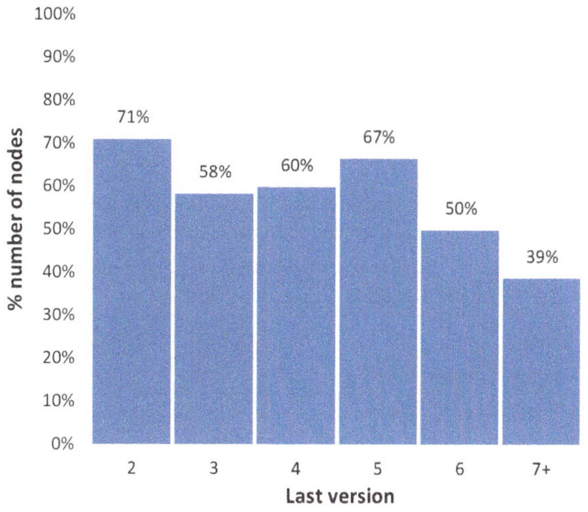

value, which presents cases that the last version location is the most accurate one, decreases—from 71% for 2 versions to 39% for 7 (and above) versions. This statistic validates our hypothesis of considering the implementation of alternative data models that use node location historical data.

4.5 Models Analysis and Evaluations

The models described in 3 are implemented for all 506 nodes, with the statistical results presented in Tables 3 and 4. We examined all node locations calculated via the different models, and the last node version in respect to the reference data. When looking at the overall map quality in terms of all 506 node position accuracy, depicted in Table 3, the last version and the cluster-based model produce very similar results, where the average difference distance in respect to the corresponding reference nodes is 3.95 and 3.90 m, respectively. The other two models produce inferior results. The RMSE value is slightly better for the cluster-based model, but with no definite significance.

Looking at Table 4, it is evident that even when simple arithmetic mean calculation is implemented ("Mean"), for 30% of all nodes the last version location is less accurate than the calculated location that relies on all existing versions per node.

Table 3 Statistics for different node location calculation models when compared to the reference: last version, mean, weighted mean, and cluster-based

Calculation	Model							
	Last Version		Mean		Weighted Mean		Cluster-based	
	Average (m)	RMSE (m)	Average (m)	RMSE (m)	Average (m)	RMSE (m)	Average (m)	RMSE (m)
Eucl. distance	**3.95**	**4.90**	6.51	8.04	5.30	6.41	**3.90**	**4.75**
dX	**−0.60**	**3.00**	−2.08	6.26	−1.73	4.81	**−0.40**	**3.00**
dY	**−1.40**	**3.90**	−2.03	5.04	−1.34	4.25	**−1.40**	**3.70**

Table 4 Statistics for different node location calculation models when compared to the reference: last version, mean, weighted mean, and cluster-based

The model	% of improvement (>2 m)	% of worsening (>2 m)	% of improvement	% of worsening
Mean	30	70	21	79
weighted mean	38	62	24	76
Cluster-based	**58**	42	**76**	24

% of improvement and worsening compared to the last version

Improvement rate is slightly higher when the "Weighted Mean" model is implemented: 38% of node locations are improved when compared to the last version. Examining the cluster-based approach, for 58% of all nodes a better location is automatically calculated with respect to the last version. 25% of all the nodes have small improvement or worsening values that are smaller than |0.4| meters (which is the accuracy value of the reference data). If we examine nodes with improvement or worsening higher than 2 m (the median value of all distances), which are 28% of the examined 506 nodes, the overall betterment percentage of the cluster-based model grows to 76% (instead of 58%).

Figure 8 depicts graphs showing the distances for all node locations—cluster-based model (blue) and last version (red)—in respect to the reference location, in ascending order (version 2 top-left to version 7 + lower-right). A red dot that is above its counterpart blue dot means that the last version location is worse than the cluster-based computed location—and vice versa. For nodes having 2 versions, a straightforward weighted mean computation produced more accurate results when compared to the last version. While for versions 3–5 there is no evident conclusion in terms of better location accuracy (roughly the same number of last version nodes above and below), versions 6 and above show that most last version node locations are above the cluster-based/weighted ones, meaning that for most nodes the implemented models automatically compute more accurate locations. Although the number of nodes having 6 and above versions is not high, still a general conclusion can be stated that the higher number of versions exist, the potential of computing a more accurate location that relies on the accumulated historical location data is increased.

5 Conclusions and Future Works

In this research, we have evaluated and analyzed the positional accuracy of OSM node entities in terms of positional accuracy and geometrical representation focusing on the historical data. This was primarily done to assess the current practice of using the last version in OSM as the most accurate location. To this end, we have introduced and implemented an alternative automatic cluster-based model calculation that uses the existing node location versions for calculating node locations.

Statistical analyses showed that in general, the more versions exist, the node positional accuracy can be improved, thus validating the correlation related to the number of contributors/updates in OSM data and the overall OSM map positional accuracy. A closer inspection showed that the location of the last version is not always the most accurate one among all other versions, especially when more than 6 versions exist. Statistical analyses showed that the alternative cluster-based model was robust in identifying clusters of node location versions, filtering node location versions that are regarded as errors and outliers. Accordingly, in our study area, for 58% of the 506 analyzed nodes the model was able to automatically calculate more accurate locations in respect to the last version. The percentage value increases to 76% if we take into consideration nodes that their location is improved or worsened

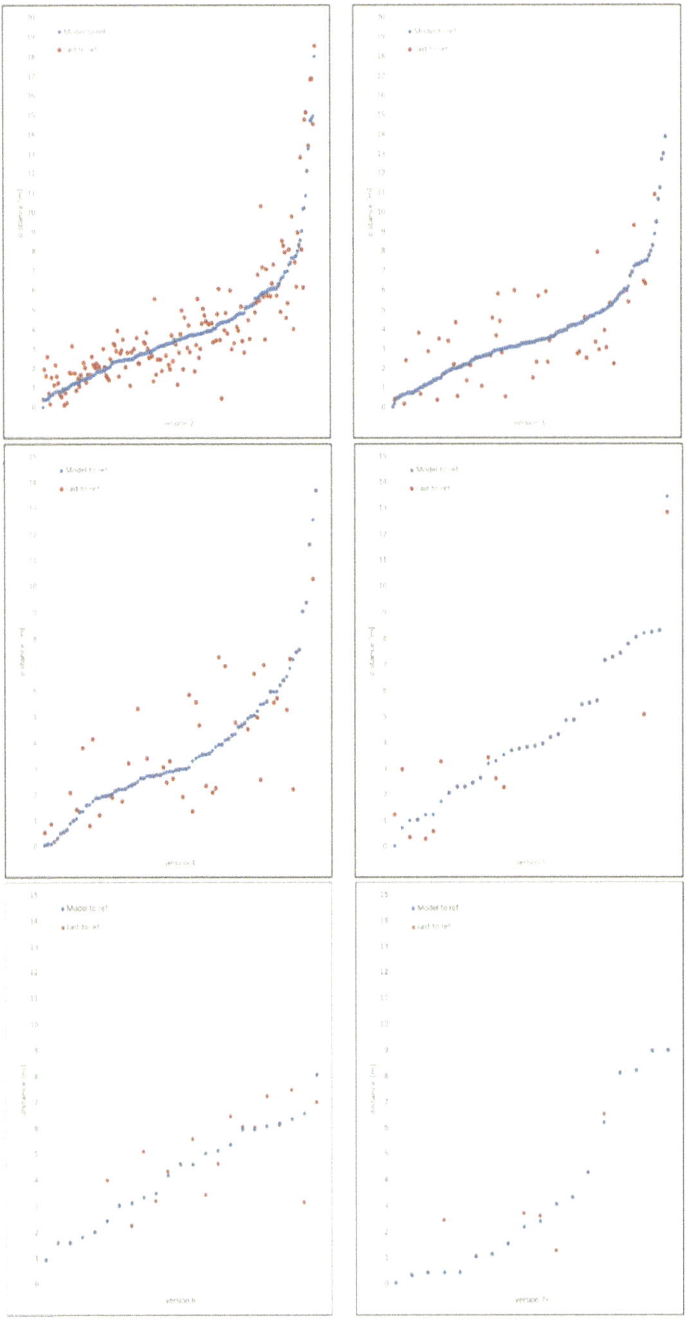

Fig. 8 Comparison to reference location: cluster-based (blue) and last version (red)

in a significant value greater than 2 m. A strong correlation was found between the number of node versions and the likelihood that the cluster-based model will compute a better node location. These results are in line with the conclusions made in Nasiri et al. [31], which showed an improvement of completeness and positional precision of OSM road features of up to 14% by exploiting OSM historical archive data.

Future research is needed to expand the case study area to other areas for examining this model on larger datasets to ascertain its robustness and scalability. This will also serve to fine-tune and adjust the parameters and thresholds used in the model. Also, we plan to improve the model by incorporating geometrical and topological constraints associated with the features represented by the nodes, which should produce better results.

We conclude that our cluster-based model for automatic node location calculation robustly handles the existence of outliers in the data by filtering erroneous observations, while using the versions that can contribute to the overall location calculation. Since one of the limitations of user-generated data is heterogeneity, implementing this model seems more appropriate for OSM rather than using the last version only, also validating the patchwork approach associated with crowdsourced user-generated geographic data. Moreover, this model uses adaptive thresholds calculated on the available data, and it does not require preliminary knowledge or predefined data used as a reference in its implementation. We believe that the developed methodology is an important step towards improving the overall positional accuracy and quality of OSM, contributing to the general motivation of using crowdsourced user-generated geographic data.

Acknowledgements The authors would like to thank Mr. Mordechai Schaap, for providing the reference data used in this research, and for his ArcGIS support for data retrieval.

References

1. O'Reilly, T.: What is web 2.0: Design patterns and business models for the next generation of software. Commun. Strat 1(65), 17–37
2. Haklay, M., Singleton, A., Parker, C.: Web mapping 2.0: The neogeography of the GeoWeb. Geogr. Compass **2**(6), 2011–2039 (2008)
3. Haklay, M.: How good is volunteered geographical information? A comparative study of OpenStreetMap and ordnance survey datasets. Environ. Plann. B: Plann. Des. **37**(4), 682–703 (2010)
4. Howe, J.: The rise of crowdsourcing. Wired Mag. **6**(14), 183–187 (2006). https://www.wired.com/wired/archive/14.06/crowds_pr.html. Last accessed 2019/10/31
5. Haklay, M., Weber, P.: OpenStreetMap: User-generated street maps. IEEE Perv. Comput. **7**(4), 12–18 (2008)
6. Siriba, D.N., Dalyot, S.: Adoption of volunteered geographic information into the formal land administration system in Kenya. Land Use Policy **63**, 279–287 (2017)
7. Neis, P., Zielstra, D.: Recent developments and future trends in volunteered geographic information research: The case of OpenStreetMap. Fut. Internet **6**(1), 76–106 (2014)
8. Mooney, P., Corcoran, P.: Characteristics of heavily edited objects in OpenStreetMap. Fut. Internet **4**(1), 285–305 (2012)

9. Neis, P., Zipf, A.: Analyzing the contributor activity of a volunteered geographic information project—The case of OpenStreetMap. ISPRS Int. J. Geo-Inf. **1**(2), 146–165 (2012)
10. Senaratne, H., Mobasheri, A., Ali, A.L., Capineri, C., Haklay, M.: A review of volunteered geographic information quality assessment methods. Int. J. Geogr. Inf. Sci. **31**(1), 139–167 (2017)
11. Ciepłuch, B., Mooney, P., Jacob, R., Zheng, J., Winstanely, A.C.: Assessing the quality of open spatial data for mobile location-based services research and applications. Arch. Photogram. Cartogr. Remote Sens. **22**, 105–116 (2011)
12. Ludwig, I., Voss, A., Krause-Traudes, M.: A Comparison of the Street Networks of Navteq and OSM in Germany. In: Geertman, S., Reinhardt, W., Toppen, F. (eds.) Advancing Geoinformation Science for a Changing World. Lecture Notes in Geoinformation and Cartography, pp. 65–84. Springer, Berlin (2011)
13. Neis, P., Zielstra, D., Zipf, A.: The street network evolution of crowdsourced maps: OpenStreetMap in Germany 2007–2011. Fut. Internet **4**(1), 1–21 (2011)
14. Zielstra, D., Hochmair, H.H.: Comparative study of pedestrian accessibility to transit stations using free and proprietary network data. Transp. Res. Rec. J. Transp. Res. Board **2217**(1), 145–152 (2011)
15. Arsanjani, J.J., Barron, C., Bakillah, M., Helbich, M.: Assessing the quality of OpenStreetMap contributors together with their contributions. In: 16th AGILE International Conference on Geographic Information Science, pp. 14–17. Leuven, May 2013
16. Zheng, S., Zheng, J.: Assessing the completeness and positional accuracy of OpenStreetMap in China. In: Thematic Cartography for the Society, pp. 171–189. Springer, Cham (2014)
17. Tenney, M.: Quality Evaluations on Canadian OpenStreetMap Data. Spatial Knowledge And Information. McGill University, Montreal, QC (2014)
18. Hochmair, H.H., Zielstra, D., Neis, P.: Assessing the completeness of bicycle trail and lane features in OpenStreetMap for the United States. Trans. GIS **19**(1), 63–81 (2015)
19. Brovelli, M.A., Minghini, M., Molinari, M., Mooney, P.: Towards an Automated Com-parison of OpenStreetMap with Authoritative Road Datasets. Trans. GIS **21**(2), 191–206 (2017)
20. Hristova, D., Mashhadi, A., Quattrone, G., Capra, L.: Mapping community engagement with urban crowd-sourcing. In: Sixth International AAAI Conference on Weblogs and Social Media, pp. 14–19, Dublin, Ireland, June 2012
21. Girres, J.F., Touya, G.: Quality assessment of the French OpenStreetMap dataset. Trans. GIS **14**(4), 435–459 (2010)
22. Al-Bakri, M., Fairbairn, D.: Assessing the accuracy of 'crowdsourced' data and its integration with official spatial data sets. In: Tate, N.J., Fisher, P.F. (eds.) The 9th International Symposium on Spatial Accuracy Assessment in Natural Resources and Environmental Sciences, pp. 317–320, Leicester, July (2010)
23. Helbich, M., Amelunxen, C., Neis, P., Zipf, A.: Comparative spatial analysis of positional accuracy of OpenStreetMap and proprietary geodata. In: GI_Forum 2012: Geovizualisation, Society and Learning, pp. 24–33. Salzburg, Germany (2012)
24. Jackson, S.P., Mullen, W., Agouris, P., Crooks, A., Croitoru, A., Stefanidis, A.: Assessing completeness and spatial error of features in volunteered geographic information. ISPRS Int. J. Geo-Inf. **2**(2), 507–530 (2013)
25. Haklay, M., Basiouka, S., Antoniou, V., Ather, A.: How many volunteers does it take to map an area well? The validity of Linus' law to volunteered geographic information. Cartogr. J. **47**(4), 315–322 (2010)
26. Mashhadi, A., Quattrone, G., Capra, L., Mooney, P.: On the accuracy of urban crowd-sourcing for maintaining large-scale geospatial databases. In: WikiSym 2012 Conference Proceedings - 8th Annual International Symposium on Wikis and Open Collaboration, pp. 15–24. Linz, Austria, Aug 2012
27. Corcoran, P., Mooney, P.: Characterising the metric and topological evolution of OpenStreetMap network representations. Eur. Phys. J. (EPJ): Spec. Top. **215**(1):109–122 (2013)

28. Vandecasteele, A., Devillers, R.: Improving volunteered geographic information quality using a tag recommender system: the case of OpenStreetMap. In: Jokar Arsanjani, J., Zipf, A., Mooney, P., Helbich, M. (eds.) OpenStreetMap in GIScience. Lecture Notes in Geoinformation and Cartography, pp. 59–80. Springer, Cham (2015)
29. Keßler, C., De Groot, R.T.A.: Trust as a Proxy measure for the quality of volunteered geographic information in the case of OpenStreetMap. In: Vandenbroucke, D., Bucher, B., Crompvoets, J. (eds.) Geographic information science at the heart of Europe. Lecture Notes in Geoinformation and Cartography, pp. 21–37. Springer, Cham (2013)
30. Zhao, Y., Zhou, X., Li, G., Xing, H.: A spatio-temporal VGI model considering trust-related information. ISPRS Int. J. Geo-Inf. **5**(2), 10 (2016)
31. Nasiri, A., Ali Abbaspour, R., Chehreghan, A., Jokar Arsanjani, J.: Improving the quality of citizen contributed geodata through their historical contributions: the case of the road network in OpenStreetMap. ISPRS Int. J. Geo-Inf. **7**(7), 253 (2018)
32. Barron, C., Neis, P., Zipf, A.: A Comprehensive framework for intrinsic OpenStreetMap quality analysis. Trans. GIS **18**(6), 877–895 (2014)
33. Ester, M., Kriegel, H. P., Sander, J., Xu, X.: A density-based algorithm for discovering clusters in large spatial databases with noise. In: 2nd International Conference on Knowledge Discovery and Data Mining (KDD-96), pp. 226–231, Aug 1996
34. Liu, P., Zhou, D., Wu, N.: VDBSCAN: varied density based spatial clustering of applications with noise. In: 2007 International Conference on Service Systems and Service Management, pp. 1–4. IEEE, Chengdu, China, June 2007
35. Mooney, P., Corcoran, P.: Accessing the history of objects in OpenStreetMap. In: Proceedings AGILE, pp. 155, Apr 2011
36. Araujo, A., Giné, E.: The central limit theorem for real and Banach valued random variables. Wiley, New York (1980)

Open Geospatial Data Contribution Towards Sentiment Analysis Within the Human Dimension of Smart Cities

Tiago H. Moreira de Oliveira and Marco Painho

Abstract In recent years, there is a widespread growth of smart cities. These cities aim to increase the quality of life for its citizens, making living in an urban space more attractive, livelier, and greener. In order to accomplish these goals, physical sensors are deployed throughout the city to oversee numerous features such as environmental parameters, traffic, and the resource consumption. However, this concept lacks the human dimension within an urban context, not reflecting how humans perceive their environment and the city's services. In this context there is a need to consider sentiment analysis within a smart city as a key element toward coherent decision making, since it is important not only to assess what people are doing, but also, why they are behaving in a certain way. In this sense, this work aims to assemble tools and methods that can collect, analyze and share information, based on User Generated spatial Content and Open Source Geospatial Science. The emotional states of citizens were sensed through social media data sources (Twitter), by extracting features (location, user profile information and tweet content by using the Twitter Streaming API) and applying machine learning techniques, such as natural language processing (Tweepy 3.0, Python library), text analysis and computational linguistics (Textblob, Python library). With this approach we are capable to map abstract concepts like sentiment while linking both quantitative and qualitative analysis in human geography. This work would lead to understand and evaluate the "immaterial" and emotional dimension of the city and its spatial expression, where location-based social networks, can be established as pivotal geospatial data sources revealing the pulse of the city.

Keywords Smart cities · Open geospatial data · User generated spatial content (UGsC) · Sentiment analysis · Ambient geographic information (AGI)

T. H. M. de Oliveira (✉) · M. Painho
NOVA IMS, Universidade NOVA de Lisboa, Lisbon, Portugal
e-mail: tiago.oliveira@novaims.unl.pt

M. Painho
e-mail: painho@novaims.unl.pt

A. Mobasheri (ed.), *Open Source Geospatial Science for Urban Studies*, Lecture Notes in Intelligent Transportation and Infrastructure,
https://doi.org/10.1007/978-3-030-58232-6_5

1 Introduction

Cities, as government units, are becoming increasingly larger, more complex and more important as the population ranks of urban areas swell with ever increasing speed. According to the United Nations Population Fund, 2008 marked the year when more than 50% of all people, 3.3 billion, 23 lived in urban areas. By 2030 this number is expected to increase to 5 billion [1].

In this sense, a smart city's main purpose is to increase the quality of life for its citizens and to make the city more attractive, livelier, and greener [2]. To do so, physical sensors are deployed throughout the city to monitor various features such as environmental parameters (weather, pollution, etc.), traffic, and the consumption of resources [3]. This live state, however, includes only measurable quantities and disregards how the citizens feel. It is likely that correlations exist between the emotional states of the citizens and relevant statistics like well-being of the city inhabitants or quality of living [4]. When urban planners use the collected data to optimize parts of the city, the emotional state of the inhabitants can thus serve as valuable implicit feedback.

This gives rise to the vision of an emotion-aware city with the ability to understand and utilize the emotional states of its citizens to enable improved planning and decision making, in which urban citizens can be called upon to act as active sensors, sharing their spatiality. Therefore, the use of User Generated spatial Content (UGsC), including social media information, could truly lead to better urban planning [5–7].

This chapter aims to deliver a literature review regarding the definition of a smart city and the importance of its human/emotional component, while presenting a methodological approach towards sentiment analysis and mapping, to assess people's emotional response to their environment. This methodology, based in an Ambient Geographic Information (AGI) approach, gathers open source geospatial data through User Generated spatial Content (UGsC) retrieved from Social Media (Twitter) related with people's perception and feelings, and therefore characterize its emotional dimension.

This work is structured in the following sections: the current section will provide some insights regarding the definition of smart cities and concepts such as crowdsourcing, UGsC and AGI; Sect. 2 will address issues regarding emotion and sentiment analysis and its relevance smart cities; on Sect. 3, the methodological approach to harvest tweets with sentiment content will be presented; on Sect. 4, some preliminary results will be presented; finally, the last section will present some of the possible outcomes of the research along with proposed future work.

1.1 In the Pursuit of Smart Cities

Nowadays, since there is a rapid increase of urban population worldwide, cities face a variety of risks, concerns, and problems, namely: physical risks such as deteriorating conditions in air and transportation, and economic risks such as unemployment [1]. This unprecedented rate of urban growth creates an urgency to find new approaches and methods to tackle these challenges. Smart cities could be one of the reactions for these issues since they aim to improve "urban functions" and services provided to the population not only through innovation, but also through the combination of networks (especially wireless), sensors/actuators and the active commitment of citizens [8].

But how exactly can we define them? Table 1 presents some working definitions of what is classified as a smart city. While some authors [9] and focus on the use of smart computing technologies, others [10] highlight the performance of a smart city in economy, people, governance, mobility, environment, and living. In a Project of the Natural Resources Defense Council [11] the emphasis is given to the positive outcomes made by being "smarter". However, most of the definitions stress technologies [12–14], in which the city is dependent by the sourcing of real-time real-world

Table 1 Smart city definitions (Author)

Author	Definition	Smart city dimension focus
Washburn, D. (2009) [9]	*The use of Smart Computing technologies to make the critical infrastructure components and services of a city—which include city administration, education, healthcare, public safety, real estate, transportation, and utilities—more intelligent, interconnected, and efficient*	Use of smart computing technologies
Giffinger, R. (2007) [10]	*A city well performing in a forward-looking way in economy, people, governance, mobility, environment and activities of self-decisive, independent and aware citizens*	Smart city impacts
Appleyard, B. (2007) [11]	*A city striving to make itself "smarter" (more efficient, sustainable, equitable, and livable)*	Smart city impacts
Hall, R. (2000) [12]	*A city that monitors and integrates conditions of all of its critical infrastructures, including roads, bridges, tunnels, rails, subways, airports, seaports, communications, water, power, even major buildings, can better optimize its resources, plan its preventive maintenance activities, and monitor security aspects while maximizing services to its citizens*	Technology

(continued)

Table 1 (continued)

Author	Definition	Smart city dimension focus
Harrison, C. (2010) [13]	*An instrumented, interconnected, and intelligent city. Instrumentation enables the capture and integration of live real-world data through the sensors, kiosk, meters, personal devices, appliances, cameras, smart phones, implanted medical devices, the web, and other similar data-acquisition systems, including social networks as networks of human sensors. Interconnected means the integration of those data into an enterprise computing platform and the communication of such information among the various city services. Intelligent refers to the inclusion of complex analytics, modeling, optimization, and visualization in the operational business processes to make better operational decision*	Technology
Rios, P. (2012) [14]	*A city that gives inspiration, shares culture, knowledge, and life, a city that motivates its inhabitants to create and flourish in their own lives*	Technology
Partridge, H. (2004) [15]	*A city where the ICT strengthen the freedom of speech and the accessibility to public information and services*	Human dimension

data from both physical and virtual sensors. Such data may be interconnected across multiple processes, systems, organizations, industries, or value chains. The combination of instrumented and interconnected systems effectively connects the physical world to the virtual world. Few authors [15] shed light on the human dimension of a smart city.

There are four dimensions in which a smart city primarily operates, namely: intelligent city (its social infrastructure), the digital city (informational infrastructure), the open city (open governance) and the live city (a continuously adaptive urban living fabric) [8]).

Figure 1 identifies and clarifies the main conceptual variants and core factors of a smart city according to Nam and Pardo [1]: Technology (infrastructures of hardware and software); People (creativity, diversity, and education); and Institution (governance and policy).

The above mentioned human component of the smart city relates to the idea that a smart city is also a living urban fabric that is continuously being reshaped as is

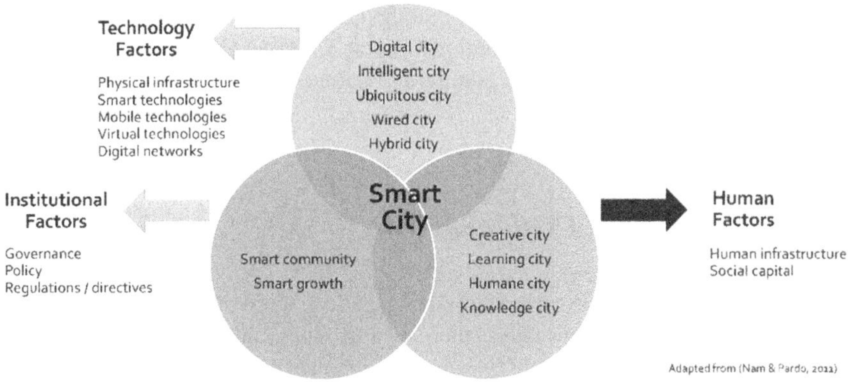

Fig. 1 Fundamental components of a smart city

adaptive to change [8]. Essentially, urban functions should not be disconnected from urban planning, and its immaterial dimensions, such as security perception, sense of belonging and joy, in which could affect positively or negatively health and urban population well-being [16].

In this context, there's a need to consider affect and emotion within a smart city, as a key element towards rational decision making [17]. Since emotion is a central component of human behavior and, for a city to be truly "smart", it is important not only to assess what people are doing, but also, why they are behaving in a certain way [18], as considering emotional states is essential for achieving real-time judgment and perceived life satisfaction [19]. With this information, city planners can make use of the gathered affective data to detect positive or negative trends developing in the city, managing to take early countermeasures. Answering subjective questions such as "which part of the city is the best?", requires affect and emotion [19].

But more important for a smart city is its capability to capture the sense of places. A city is not a machine, but rather made by people local actions and feelings. This could not be captured and represented without active citizens sensors (Volunteered Geographic Information [20], Ambient Geographic Information [21], crowdsourcing [22] connected to location based-social networks [23]. Due to the relative pre-eminence of place to understand urban environment, and following Professor Michael Goodchild insights [24], it is time to move from a space-based to a place-based geospatial infrastructure.

Smart urban solutions should be built on the vision of citizens as active sensors on one hand, and on the other hand on spatial enablement of citizens via social network. These kinds of solutions have also to be built on the potentials offered both by embedded sensors to crowdsource the process of collecting geo-referenced information about places in the city. These constructs gave rise to the vision of an emotion-aware city, in which the "immaterial" (and human) dimension is the main component, and the main focus is given on the need that smart cities have to assess their citizen's feelings, perception and well-being [25]. An emotion-aware city can be defined by its capacity on interpreting and harnessing the affective states of its citizens [26].

1.2 Crowdsourcing, UGsC and AGI

As explained by Roche [23], an active and engaged citizen is indeed the main driving force of a "smart city". Nowadays, there is a growing amount of location-based contents generated by connected "*produsers*", mainly equipped with smartphones. The exponential growth of ambient geographic information through social networks became the basic feature of a spatially enabled society, in which it behaves as a vessel where millions of people share their current thoughts, observations and opinions, showing to provide more reliable and trustworthy information than traditional methods like questionnaires and other sources [27]. A spatially enabled citizen is explained through his ability to express, formalize, equip (technologically and cognitively), and (un)consciously activate an efficiently use of his spatial skills [23].

Social media generated from many individuals plays a greater role in our daily lives while providing a unique opportunity to gain valuable insight on information flow and social networking within a society [28]. Through data collection and analysis of its content, it supports mapping and understanding of the evolving human landscape [21]. In this context, mobile technologies are definitely a valuable tool for collecting affective data in the context of an emotion-aware city [29], since they can simultaneously collect both spatial location and the user posts, which should contain emotion or mood content.

Harvesting this ambient geospatial information provides a unique opportunity to gain valuable insight on information flow and social networking within a society, support a greater mapping, understand the human landscape and its evolution over time [21]. This emergence of AGI represents a second step in the evolution of geospatial data availability, following on the heels of Volunteered Geographic Information (VGI) [30], in which harvesting and analyzing such ambient information represents a substantial challenge, needing new skillsets as it resides at the intersection of disciplines like geography, computational social sciences, linguistics and computer science [21].

Ambient Geographic Information (AGI) can be defined as geographic information that can be harvested from social media feeds. Since VGI is primarily about crowdsourcing with specific tasks outsourced to the public at large, AGI is different, since it focuses on "crowd-harvesting", with the general public broadcasting information that can be harvested in a meaningful manner [30]. Given that, the emergence of social media participation and information sharing is bringing forward a different model of geospatial information contribution, meaning that nowadays users' intent is not to directly contribute geospatial data (e.g. a map), but rather to contribute information (e.g. a geotagged picture from a family vacation, or text describing a planned event) that happens to have an associated geospatial component and a corresponding footprint.

Assembling and analyzing AGI provide us with unparalleled insight on a broad variety of cultural, societal, and human factors, particularly as they relate to human and social dynamics, for example: (1) mapping the manner in which ideas and information propagate in a society, information that can be used to identify appropriate strategies for information dissemination during a crisis situation. (2) Mapping people's opinions and reaction on specific topics and current events, thus improving our ability to collect precise cultural, political, economic and health data, and to do so at near real-time rates. (3) Identifying emerging socio-cultural hotspots.

Distinct from VGI where people are acting *only* as sensors, in AGI they are also the observations from which we can get a better understanding of various parameters of the human landscape. On the other hand, AGI focuses upon passively contributed data, unlike VGI.

Unfortunately, the geolocation features offered by social networks are not very commonly used, which generates a great variation in the availability of geolocation information. As an example, the most popular social networks, such as Instagram, Twitter and Facebook, do not have much posts with location attached to them, with respectively 30%, 15% and 4% [31].

Since social media has rose in popularity and scale, there is a growing need to extract useful information from huge amounts of data. Social networks like Twitter, with an estimated community of 332 million worldwide and more than 400 million posts every day [32], can potentially serve as a valuable information resource for various applications.

There are some case studies in which social media contributions lead to perceptions and images of real-world phenomena. Since twitter contains a large corpus of public real-time data that, it has been used successfully in studies regarding disasters and emergencies [33], public health monitoring [34], events exploration [35], among others.

The rise of social media and the ability for analysis raises several concerns with respect to the suitability of traditional mapping and GIS solutions to handle this type of information [22, 36]. It is now possible to map abstract concepts like the flow of information in a society, contextual information to place and linking both quantitative and qualitative analysis in human geography [21].

In a sense one could consider AGI to be addressing the fact that the human social system is a constantly evolving complex organism where people's roles and activities are adapting to changing conditions and affect events in space and time. By moving beyond simple mashups of social media feeds to actual analysis of their content we gain valuable insight into this complex system [21].

2 Sentiment Analysis

Mapping emotion builds on a tradition of studies in cognitive mapping, evaluative mapping, environmental preference and environmental affect [37], adding an approach in which people experience, evaluate, and describe their environment in situ through social media.

As stated before, one of the most easily accessible public data sources that can be used to detect emotion are social networks. People use them to share their opinions and/or express their feelings. It has been shown that users of social media tend to post authentic and reliable information about themselves [27]. This type of approach brings an opportunity to detect sentiment.

Approaches based on text analysis are suited for extracting moods and emotions from social networks, since linguistic characteristics analysis on the written posts made by individuals on their pages can be used to infer negative and positive moods [38]. By applying this technique to Twitter/Facebook/Flickr/Instagram update messages allows to reveal information about public moods and emotions [39], additionally Twitter "hashtags" can be harnessed to extract individual mood states [40].

As explained by Dramstad [41], most people, if questioned, will have an opinion as to whether a landscape is aesthetically pleasing or not, and how everyday landscapes reflect in the well-being of people is receiving increased focus in research. On the other hand, Weinreb [37] advocates that personal associations are a primary example of intangible and subjective feelings, related much more to memory than to anything immediately visual. Positive personal associations stemmed from memories about a range of personal experiences with friends and family and attachment to the place or locality. Negative personal associations can be articulated as disappointment regarding the ruined or unrealized potential of a space, often tied to a sentiment that municipal leaders could failed to follow through on promises to complete development, livability, or beautification projects.

A mental map refers to one person's point of view perception of their own world, and is influenced by that person's culture, background, mood and emotional state, instantaneous goals and objectives [31]. For instance, if we move along the streets of any city in a rush, trying to find a certain type of shop or building, our experience will be different than the one we would have had if we were searching for something else. There are examples [42] in which a concept of a pedestrian routing system was developed, who enabled users to consider factors and quantify their influences on the extraction of walking routes, factors such as street length, greenness, sociability, and quietness.

There is a multitude of reasons why a pedestrian may choose to avoid areas with negative affect: emotions like fear and anger indicate danger and could be avoided by travelers feeling afraid. Stress may be felt in areas of high traffic or crowdedness which are undesirable for pedestrians. The emotion disgust indicates places that are unsuitable for relaxation. Likewise, there are reasons to seek areas with positive affect: a detour through a relaxing area can be acceptable when someone is feeling stressed. The emotion relaxed may also be correlated with higher safety to walk in an area. Furthermore, someone may want to seek locations with increased surprise when they are curious and, in the mood, to explore. Reasons to seek or avoid areas with a certain affective state are as manifold and personal as affect itself [43].

Likewise, a person's socio-economic status, cultural ties, and past experiences influence how people perceive environmental quality. In the case of tourism, people using these areas can differ in many ways, including their personal characteristics and perception about the recreation environment [44].

Since urban planning is a process of regulating land use to optimize aspects like resource consumption, transportation and safety in the face of rapid urban growth, negative trends in the city should be recognized as early as possible, by using citizens as sensors [4, 45]. If problems are identified quickly, early countermeasures can be taken. For instance, there are problems that are obvious to inhabitants, but are hard to measure with statistics and facts. For instance, even without available crime records, people have a good intuition of the safety of a neighborhood. Similarly, the amount of visual degradation and littering is hard to measure but can be judged subjectively and trigger an emotion. These two examples state that affective states can be harnessed to understand how citizens truly perceive their environment. These subjective feelings may differ greatly from measurable statistics.

3 Materials and Methods

In this chapter we introduce the practical component of this study in which the main purpose is to present an AGI-based methodology, designed towards sentiment analysis, which aims to create tools and methods that can collect, analyze and share information, based on Open Source Geospatial technology and User Generated spatial Content (UGsC) linked with social networks and media.

3.1 Twitter as an UGsC Data Source

For this study, the data is harvested through Twitter, a social media service where both mobile and web users post and read short messages of up to 280 characters, called tweets, that can contain links and media content, such as pictures and videos [46]. Twitter is a popular microblogging service, with more than 332 million monthly active users, 400 million tweets sent per day and 80% of users using mobile platforms [32]. The introduction of the hashtag "#" syntax facilitates discussion on a specific topic and offers an important filter system for a specific subject [47]. Tweets can have both primary and secondary status, the first when the user posts a message and the second when the tweet is a reply to another user, or a retweet broadcasting of a prior message written by another user [48].

Fig. 2 Example of a geolocated tweet with precise coordinates and textual description

Usually, geolocation information in tweets can be provided directly by the contributing bloggers, if they decide to make this information available, or it can be deduced from IP addresses using any of the IP geolocation solutions [49].

For our approach we focus on geolocation information either contributed directly by the user or provided through the client application. This geolocation information is available through precise coordinates (as shown on the upper section of Fig. 2), or in a descriptive manner (similar as the example on the lower section of Fig. 2).

On the first example, at the top, we see the tweet as it appears on a follower's stream, and this is the Twitter message that is displayed in the browser. At the bottom we see highlights from the information retrieved through the search API for this particular tweet, including its geolocation information (marked by the box) in the form of precise coordinates. The coordinates correspond to a location in the Fairview borough of New Jersey. The second example contains the descriptive geolocation information recovered for another tweet similar to the one presented previously.

The study site for this work is continental/mainland Portugal, located on the southwestern side of Europe, from which geolocated tweets were harvested.

3.2 *Open Source Geospatial Methodology*

To harvest information from social media feeds is essentially a web-mining process. Mainly, it entails in general three operations [50]: extracting data from the data providers (various social media servers) via application programming interfaces

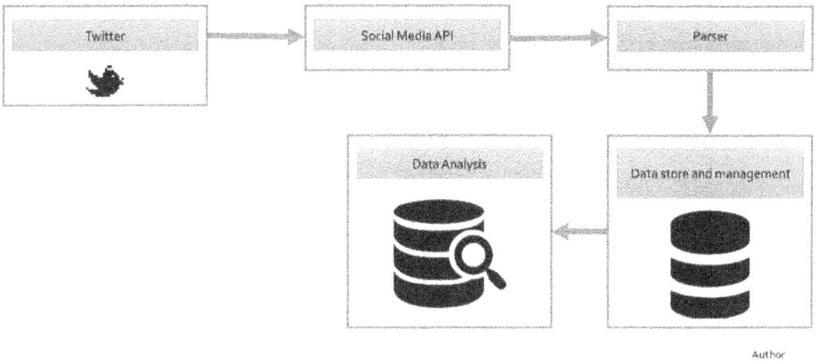

Fig. 3 General system architecture to harvest information from social media

(APIs); parsing, integrating, and storing these data in a resident database; and then analyzing these data to extract information of interest.

Data parsed from diverse sources are integrated by the parser, which organizes data from diverse sources (Fig. 3) into common categories like time of submission, username, originating location, keywords, as well as service-specific information (e.g. content, links to actual files). It allows us to establish an integrated multi-source local database that can be used to perform analysis of the harvested data that is not supported by the provider database interface (e.g. statistics on user activities) for various use cases.

The used methodological approach is as represented below in Fig. 4 and will be explained through the present section. The emotional states of citizens are sensed using Twitter, by extracting features (location, user information, posts, pictures and videos) and applying machine learning techniques, such as natural language processing, text analysis and computational linguistics.

The used method to extract tweets was through the Tweepy Python Library 3.0, which uses the Twitter Streaming API. This Open Source Python Library permits the extraction of several information related with the tweet, such as location (though precise coordinates and/or placenames/description), user profile information and the tweet content itself. The developed twitter stream service is enabled through a custom Python script made by the authors in which all the geolocated tweets from the study area are dynamically stored on a .JSON file format within a dedicated technological infrastructure with virtually zero downtime.

On our approach firstly a set of un-structured text documents is collected. Then, the pre-processing for the documents is performed to remove noise and commonly used words, stop words, stemming. This process produces a structured representation of the documents known as Term- document matrix, in which, every column represents a document and every row represents a term occurrence throughout the document. The final step is applying data mining techniques such as clustering, classification, association rules to discover term associations and patterns in the text and then, finally, visualizing these patterns using Geographic Information System tools.

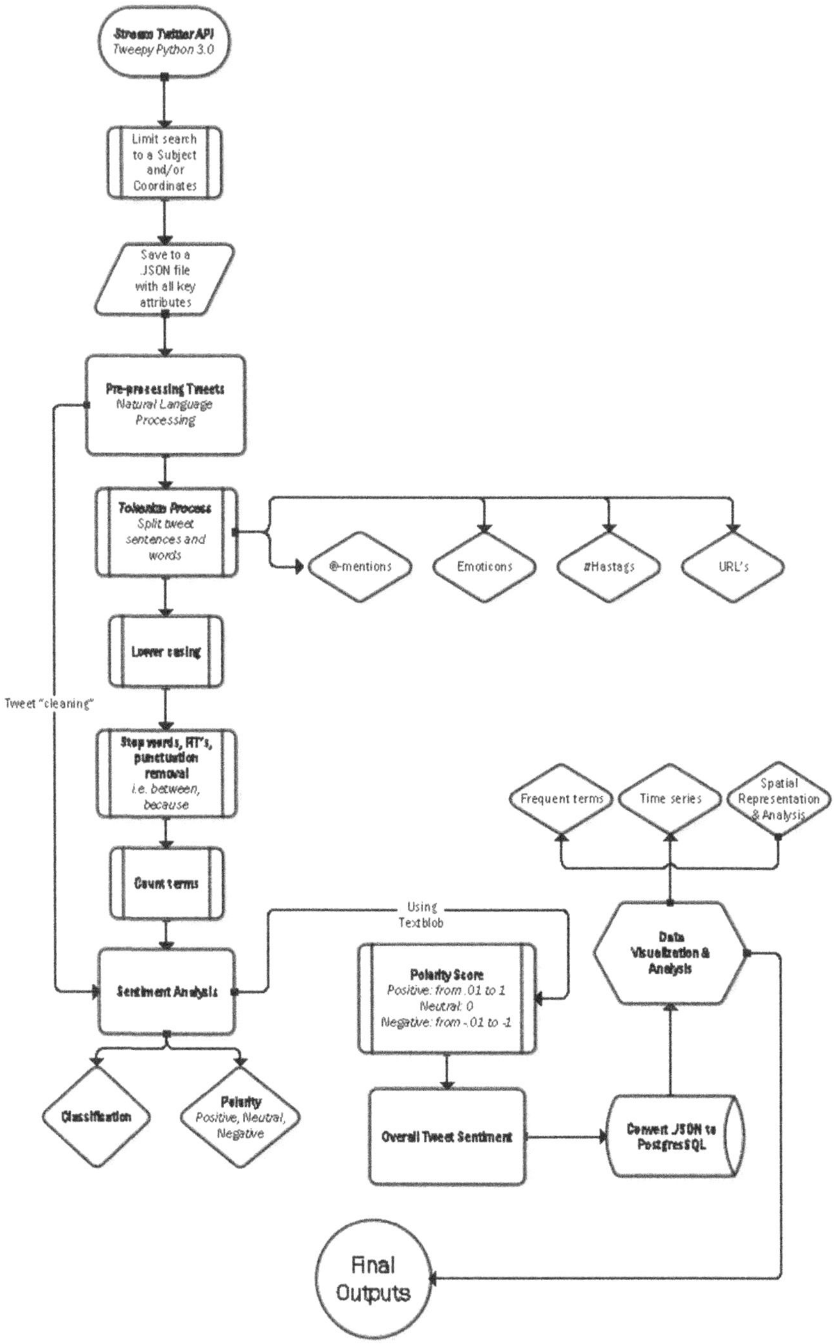

Fig. 4 Tweet harvesting and analysis methodological approach

According to Younis [32], text Mining is defined as the automated process of detecting and revealing new, uncovered knowledge and inter-relationships and patterns in unstructured textual data resources.

In this sense, and before the tweets can be used as training data, they need to be pre-processed. Since tweets are short text messages that are no longer than 280 characters and are often written on mobile devices, they typically contain many abbreviations, colloquial expressions and non-standard words [19]. Additionally, the language employed in Social Media sites is different from the one found in mainstream media and the form of the words employed is sometimes not the one we may find in a dictionary. Further on, users of Social Media platforms employ a special "slang" (i.e. informal language, with special expressions, such as "lol", "omg"), emoticons, and often emphasize words by repeating some of their letters [51]. Mainly, during the tweet pre-processing the idea is to remove the variety from the messages which constitutes noise in the training data.

The tweet pre-processing stage is comprised by the steps depicted in Fig. 5, proceeding as follows:

1. Repeated punctuation sign normalization: In the first step of the pre-processing, repetitions of punctuation signs (".", "!" and "?") are detected. Multiple consecutive punctuation signs are replaced with the labels "multistop", for the fullstops, "multiexclamation" in the case of exclamation sign and "multiquestion" for the question mark and spaces before and after.
2. Emoticon replacement: The emoticons found are replaced with their polarity ("positive", "negative" and "neutral").
3. Tokenization and lower casing process: Lower case and split the tweet content into tokens, based on spaces and punctuation signs.

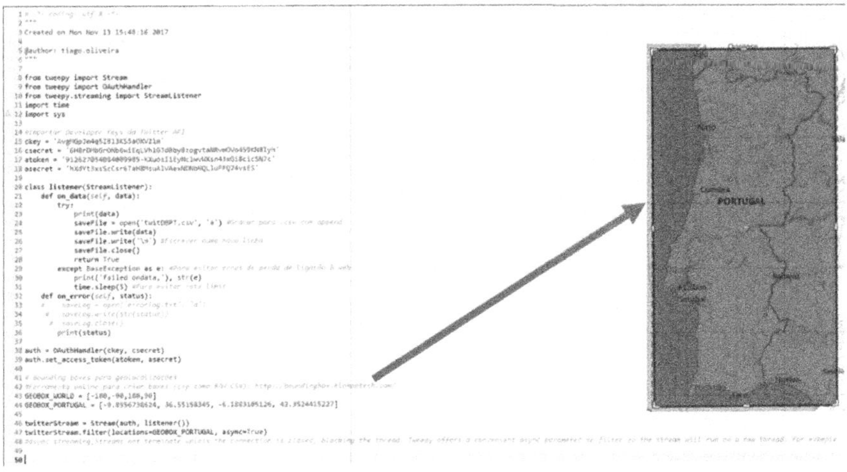

Fig. 5 Python script and study area bounding box

Balahur [51] framed sentiment analysis as the Natural Language Processing (NLP) task dealing with the detection and classification of sentiments in texts. In the literature, there is no standard method for mining and analyzing so-cial media business data [32]. On our approach, Sentiment analysis is executed with Textblob, an Open Source Python Library that processes textual data. It provides a simple API for diving into common NLP tasks such as: noun phrase extraction, part-of-speech tagging, noun phrase extraction, sentiment analysis, classification, translation, and more. In order to perform sentiment analysis with Textblob, there is the need to use also other Python libraries, namely tweepy and NLTK.

When this process is finished, the .JSON file which contains the tweets is converted to PostgresSQL database and therefore the extracted information can be used on any GIS application, both desktop and server oriented.

4 Preliminary Results and Discussion

In this chapter we present and discuss some preliminary results obtained by applying the methodology described in the previous section.

The implemented twitter streaming service so far extracted approximately 8 million georeferenced tweets, by using the below custom Python script (Fig. 5) whom harvested tweets from continental/mainland Portugal.

The harvested information was stored on a PostgresSQL database which contains three separated but interconnected tables, namely:

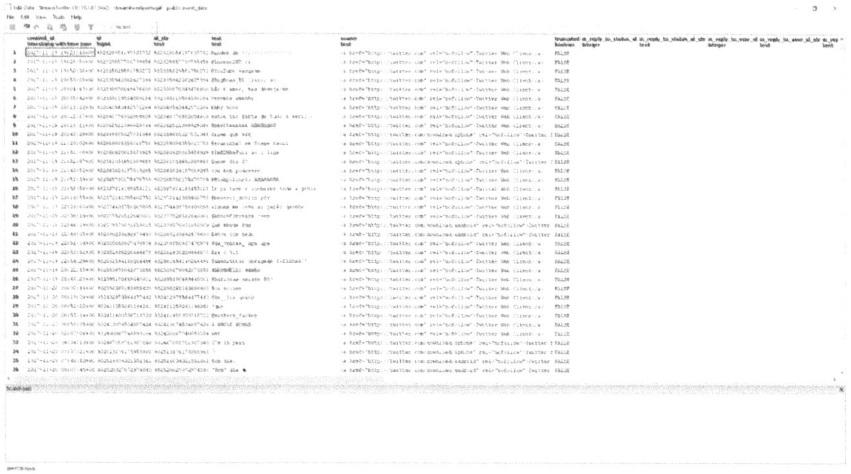

Fig. 6 Tweet data table example

1. Tweet_data (Fig. 6): which is the "parent" tweet object for the other tables, since it includes fundamental attributes such as id, created_at and text (the tweet content).
2. User_data (Fig. 7): it contains the public Twitter account metadata and describes the user which posted the tweet.
3. Place_data (Fig. 7): includes the tweet geo-tag. Tweet location can be an exact point location, or a Twitter place with a specific bounding box whose describes a larger area ranging from a venue to an entire region (Fig. 8).

Fig. 7 User data table example

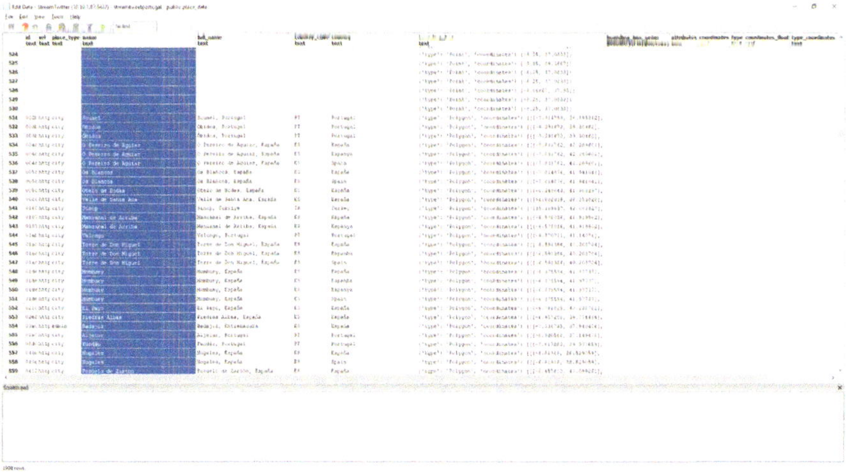

Fig. 8 Place data table example

Sentiment Analysis	
Text	If that is not cool enough for you than that is a you problem.
Polarity	-0.0875
Subjectivity	0.575
Classification	neg
P_Pos	0.344455873
P_Neg	0.655544127

Fig. 9 Sentiment analysis score example

As stated previously, with the use of NLP tools, such as Textblob (Python) on our methodology there was possible to perform sentiment analysis to each harvested tweet in which subjective information was identified and extracted. By analyzing the language used in the text, our custom Python script classified its polarity, measuring the negativity, neutralness, or the positivity of the tweet text/content, while giving it a score (from −0.01 to 1) representing if the tweet/post is negative, neutral or positive, as demonstrated on Fig. 9.

However, in Twitter, sometimes people use hashtags to notify others of the emotions associated with the message they are tweeting. However, tweets as the one showed on the fourth example (Table 2), prove that sometimes reading just the message before the hashtag does not convey the emotions of the tweeter. Here, the hashtag provides information not present (implicitly or explicitly) in the rest of the message.

On the other hand, there are also tweets, such as those shown in examples 5 and 7 that do not seem to express the emotions stated in the hashtags. This may occur for many reasons including the use of sarcasm or irony, which Textblob is not able to detect. Additional context is required to understand the full emotional impact of many tweets. Tweets tend to be very short, and often have spelling mistakes, short forms, and various other properties that make such text difficult to process by

Table 2 Smart city definitions (Author)

Tweet examples
1. RIP Ziggy Bowie Stardust…#sadness
2. My amazing memory saves the day again! #joy
3. Just saw the most awful dress ever! #Disgust
4. John used my photo on facebook. #anger
5. School is very boring today:/ #joy
6. Hope I get to school in time… #fear
7. Great morning stuck in traffic with a flat tire #love
8. I love when I pick up my car and it’s out of gas #surprise

natural language systems. Further, it is as well probable, that only a small portion of emotional tweets are hashtagged with emotion words [52].

5 Conclusions and Future Research Directions

Sentiment Analysis allows researchers to visually discern areas of strong feelings, either good or bad. This multidisciplinary approach can exhibit aggregations of positive ratings, negative ratings, or in some cases, a mixture of strong positive and negative ratings in the same place. Some other possible outputs are related with the identification of emotional patterns—those spaces where, at a specific or recurring time, a certain emotion is expressed powerfully and abundantly. It can lead us to some questions: do emotional landmarks change over time? Do they change according to the observer? To language? To the time of day, week, month or year? Additionally, several profiles of users can be established, based upon social-demographic characteristics, such as: gender; age; education level; motive of trip (leisure, business, e.g.); "level of acquaintance" of the place (old-timer, new-comer, tourist); origin (country and/or city).

This *emotional sensing* can be directed towards any topic or subject, such as: emergency scenarios such as revolts, riots or natural disasters; art and tourism; city planning and safety; entertainment and consumption, proving its relevance and replicability towards any other issue or thematic.

Contributing to cognitive mapping, emotional maps enable researchers to share a participant's position and views of the landscape as he or she articulates emotions and memories related to those views. Replicable in any setting, this technique could be used to create and maintain spaces that are attractive, inviting, and emotionally pleasing to a variety of users. These geospatial practices could highlight how emotions, subjectivities and spaces are mutually constitutive in particular places and at particular times, suggesting that people's shared feelings about specific places are influenced by the particular physical properties and characteristics of a given place, since this technique could be used to create and maintain spaces that are attractive, inviting and emotionally pleasing to a variety of users.

In this context the power of harvesting Open Source Geospatial data and UGsC stems from gaining a deeper understanding of groups rather than looking at specific individuals. As the popularity of social media is growing exponentially we are presented with unique opportunities to identify and understand information dissemination mechanisms and patterns of activity in both the geographical and social dimensions, allowing us to optimize responses to specific events, while the identification of hotspot emergence helps us allocate resources to meet forthcoming needs.

By engaging an emotion-aware city, new forms of communication can be generated. Traditionally, the choice of partners for online group communication is either based on pre-existing relationships or on similar interests or location [52]. In an emotion-aware city, communication groups can be formed spontaneously, based not only on a topic, but also on location and matching emotional state. These types of interactions can start interesting discussions about controversial projects or places within the city, since personal and sensitive issues are best shared with those who fell the same about a specific area of the city, creating participatory movements.

However, uncertainty over the data quality of UGsC (and VGI in general), is expected to be faced in this work. While using user generated information is common to struggle with positional accuracy, thematic accuracy, completeness, temporal quality and logical consistency [53, 54]. On this work it is expected to face some of those.

Besides data quality assessment, this work aims to be improved with the following developments:

1. Implementation of visual outputs of the gathered information and analysis, through maps, by producing alternative representations of space based on individuals' georeferenced experiences, thoughts and emotion;
2. Add other social media sources—as Facebook, Flickr and Instagram—and evaluate their potential as *emotional data sources*;
3. Analyze the possibility of replicate the same proposed methodology in other languages, besides English UGsCs;
4. How to understand and interpret the use of sarcasm or irony in tweets, as exemplified in Table 2;
5. Compare information retrieved from UGsC (subjective observations) with experimental data (objective measurements, such as socio-demographic statistics about a specific city), evaluating which can truly characterize and share the emotional dimension of the city;
6. Finally, to assess if there is a strong correlation between touristic sites and the *emotional landmarks* within the city, which can be defined as *emotional hot spots*.

Lastly, and revisiting the live city dimension of the smart city, it can greatly benefit from Open Source Geospatial Science, particularly from UGsC. The physical and *senseable* structure of a smart city can be analyzed through UGsC, since it provides innovative, creative, deliberative, uncertain, multi-actor, multi-scale, and multi-thematic methods and tools [25, 55].

An emotion and sentiment mapping methodology could lead to understand, assess and evaluate the *immaterial* and emotional dimensions of the city and its spatial expression. Open Source Geospatial Science can truly support the development of the intelligent city [56], due to crowdsourcing, Volunteered Geographic Information (VGI), Ambient Geographic Information (AGI), including location-based social networks which stand out as key geospatial data sources indicative of the pulse of the city.

References

1. Nam, T., Pardo, T.A.: Conceptualizing smart city with dimensions of technology, people, and institutions. The Proceedings of the 12th Annual International Conference on Digital Government Research.Challenging Times—dg.o '11, p 282 (2011)
2. Kehoe, M., Nesbitt, P.: Smarter cities series : a foundation for understanding IBM smarter cities. IBM J. Res. Dev. (2010)
3. Caragliu, A., Del Bo, C., Nijkamp, P.: Smart Cities in Europe. J. Urban Technol. **18**(2), 65–82 (2011)
4. Lathia, N., Quercia, D., Crowcroft, J.: The hidden image of the city: sensing community well-being from urban mobility. Pervasive Comput., 1–8 (2012)
5. Resch, B., Summa, A., Sagl, G., Zeile, P., Exner, J.: Urban emotions—geo-semantic emotion extraction from technical sensors, human sensors and crowdsourced data. In: Gartner, G., Haosheng, H. (eds.) Progress in Location-Based Services, pp. 199–212. Springer International Publishing, Switzerland (2014)
6. Exner, J.: Smart planning & smart cities. In: Corp 2014, vol. 8, pp. 603–610
7. Exner, J., Zeile, P., Streich, B.: Monitoring laboratory spatial planning: new benefits and potentials for urban planning through the use of urban sensing, geo- and mobile web. In: 12th International Conference on Computers in Urban Planning and Urban Management (CUPUM), pp. 1–18 (2011)
8. Roche, S.: Geographic information science I: Why does a smart city need to be spatially enabled? Prog. Hum. Geogr. **38**(5), 0309132513517365-, Feb 2014
9. Washburn, D., Sindhu, U.: Helping CIOs Understand 'Smart City' Initiatives. Massachusetts , Cambridge (2009)
10. Giffinger, R.: Smart cities ranking of European medium-sized cities, Vienna, Austria (2007)
11. Appleyard, B., et al.: Smart Cities: Solutions for China's Rapid Urbanization, New York (2007)
12. Hall, R.E., Bowerman, B., Braverman, J., Taylor, J., Todosow, H.: The vision of a smart city. In: 2nd International Life Extension Technology Workshop, p. 7 (2000)
13. Harrison, C., et al.: Foundations for smarter cities. IBM J. Res. Dev. **54**(4), 1–16 (2010)
14. Rios, P.: Creating 'The Smart City'. University of Detroit Mercy (2012)
15. Partridge, H.: Developing a human perspective to the digital divide in the smart city. In: ALIA 2004 Biennial Conference, Challenging Ideas, p. 7 (2004)
16. Schipperijn, J., Stigsdotter, U.K., Randrup, T.B., Troelsen, J.: Influences on the use of urban green space—a case study in dense, Denmark. Urban For. Urban Green. **9**(1), 25–32 (2010)
17. Naqvi, N., Shiv, B., Bechara, A.: The role of emotion in decision making a cognitive neuroscience perspective. Curr. Dir. Psychol. Sci. **15**(5) (2006)
18. Dolan, R.J.: Emotion, cognition, and behavior. Science **298**(5596), 1191–1194 (2002)
19. Guthier, B., Alharthi, R., Abaalkhail, R., El Saddik, A.: Detection and visualization of emotions in an affect-aware city. In: Proceedings of the 1st International Workshop on Emerging Multimedia Applications and Services for Smart Cities—EMASC '14, pp. 23–28 (2014)
20. Goodchild, M.F.: Citizens as sensors: The world of volunteered geography. GeoJournal **69**(4), 211–221 (2007)
21. Stefanidis, A., Crooks, A., Radzikowski, J.: Harvesting ambient geospatial information from social media feeds. GeoJournal **78**(2), 319–338 (2013)
22. Goodchild, M.F.: Twenty years of progress: GIScience in 2010. J. Spat. Inf. Sci. **1**(1), 3–20 (2010)
23. Roche, S.: Sensing places' life to make city smarter. In: ACM SIGKDD International Workshop on Urban Computing (UrbComp 2012), 12 Aug 2012, Beijing, China (2012)
24. Goodchild MF (2012) Reflections and visions, final keynote presentation. In: Global Geospatial Conference, pp. 1–22 (2012)
25. De Oliveira, T.H.M., Painho, M.: Using ambient geographic information (AGI) in order to understand emotion & stress within smart cities. In: Proceedings of AGILE 2015—18th AGILE International Conference on Geographic Information Science (2015)

26. De Oliveira, T.H.M., Painho, M.: Emotion & stress mapping—assembling an ambient geographic information-based methodology in order to understand smart cities. In: Atas da 10ª Conferência Ibérica de Sistemas e Tecnologias de Informação, vol. II, pp. 2–5 (2015)
27. Marwick AE, Boyd D (2010) I tweet honestly, I tweet passionately: Twitter users, context collapse, and the imagined audience. New Media Soc 13(1), 114–133 (2010)
28. Kar, B., Ghose, R.: Is my information private? Geo-privacy in the world of social media. ceur-ws.org, no. Webster, pp. 2009–2012 (2010)
29. Liu, Y., Piyawongwisal, P., Handa, S., Yu, L., Xu, Y., Samuel, A.: Going beyond citizen data collection with mapster: a mobile+cloud real-time citizen science experiment. In: 2011 IEEE Seventh International Conference on e-Science Work, pp 1–6 (2011)
30. Crooks, A., Croitoru, A., Stefanidis, A., Radzikowski, J.: #Earthquake: Twitter as a Distributed Sensor System. Trans. GIS **17**(1), 124–147 (2013)
31. Iaconesi, S., Persico, O.: An emotional compass harvesting geo-located emotional states from user generated content on social networks and using them to create a novel experience of cities. In: Proceedings of First International Workshop on Emotion and Sentiment in Social and Expressive Media approaches Perspectives From AI (ESSEM 2013) A Work. International Conference of the Italian Association for Artificial Intelligence, vol. 1096 (2013)
32. Younis, E.M.G.: Sentiment analysis and text mining for social media microblogs using open source tools : an empirical study. Int. J. Comput. Appl. **112**(5), 44–48 (2015)
33. Hossmann, T., Legendre, F.: Twitter in disaster mode: opportunistic communication and distribution of sensor data in emergencies. In: Proceedings of the 3rdExtreme Conference on Communication, No. 1, pp. 1–6 (2011)
34. Aramaki, E., Maskawa, S., Morita, M.: Twitter catches the flu : detecting influenza epidemics using Twitter. In: EMNLP '11 Proceedings of Conference on Empirical Methods in Natural Language Process, pp. 1568–1576 (2011)
35. Marcus, A., Bernstein, M., Badar, O.: Twitinfo: aggregating and visualizing microblogs for event exploration. In: Proceedings of SIGCHI Conference Human Factors in Computing Systems (2011)
36. Sui, D., Goodchild, M.: The convergence of GIS and social media: challenges for GIScience. Int. J. Geogr. Inf. Sci. **25**(11), 1737–1748 (2011)
37. Weinreb, A.R.: Mapping feeling: an approach to the study of emotional response to the built environment and landscape. J. Archit. Plann. Res. 1–19 (2013)
38. Leshed, G., Kaye, J.: Understanding how bloggers feel: recognizing affect in blog posts. In: CHI'06 Extended Abstracts on Human Factors in Computing Systems, pp. 2–7 (2006)
39. Bollen, J., Mao, H., Pepe, A.: Modeling public mood and emotion: Twitter sentiment and socio-economic phenomena. ICWSM, pp. 450–453 (2011)
40. De Choudhury, M., Counts, S., Gamon, M.: Not all moods are created equal! exploring human emotional states in social media. In: The Sixth International AAAI Conference on Weblogs and Social Media, no. 1 (2012)
41. Dramstad, W.E., Tveit, M.S., Fjellstad, W.J., Fry, G.L.A.: Relationships between visual landscape preferences and map-based indicators of landscape structure. Landsc. Urban Plan. **78**(4), pp. 465–474 (2006)
42. Novack, T., Wang, Z., Zipf, A.: A system for generating customized pleasant pedestrian routes based on openstreetmap data. Sensors (Switzerland), 18(11) (2018)
43. Siemer, M., Mauss, I., Gross, J.J.: Same situation–different emotions: how appraisals shape our emotions. Emotion **7**(3), 592–600 (2007)
44. Petrosillo, I., Zurlini, G., Corlianò, M.E., Zaccarelli, N., Dadamo, M.: Tourist perception of recreational environment and management in a marine protected area. Landsc. Urban Plan. **79**(1), 29–37 (2007)
45. Zheng, Y., Liu, Y., Yuan, J., Xie, X.: Urban computing with taxicabs. In: Proceedings of 13th International Conference on Ubiquitous Computing—UbiComp '11, p. 89 (2011)
46. Reips, U.D., Garaizar, P.: Mining twitter: a source for psychological wisdom of the crowds. Behav. Res. Methods **43**(3), 635–642 (2011)

47. De Longueville, B., Smith, R., Luraschi, G.: OMG, from here, I can see the flames!: a use case of mining location based social networks to acquire spatio-temporal data on forest fires. In: Conference: Proceedings of the 2009 International Workshop on Location Based Social Networks, no. c, pp. 73–80 (2009)
48. Schneider, J., Passant, A., Groza, T., Breslin, J.G.: Argumentation 3.0: how semantic web technologies can improve argumentation. In: Computational Models of Argument Proceedings, COMMA 2010, vol. 1380 (2010)
49. Eriksson, B., Barford, P., Sommers, J., Nowak, R.: A learning-based approach for IP geolocation. Passiv. Act. Meas., 171–180 (2010)
50. Sakaki, T., Okazaki, M., Matsuo, Y.: Earthquake shakes Twitter users: real-time event detection by social sensors. In: WWW'10 Proceedings of 19th International Conference on World Wide Web, p. 851 (2010)
51. Balahur, A.: Sentiment Analysis in Social Media Texts. Wassa **2013**, 120–128 (2013)
52. Ellison, N.B., Steinfield, C., Lampe, C.: Connection strategies: social capital implications of Facebook-enabled communication practices. New Media Soc. **13**(6), 873–892 (2011)
53. Foody, G., et al.: Mapping and the Citizen Sensor, First. Ubiquity Press Ltd., London (2017)
54. Mocnik, F.B., Mobasheri, A., Griesbaum, L., Eckle, M., Jacobs, C., Klonner, C.: A grounding-based ontology of data quality measures. J. Spat. Inf. Sci. **16**(16), 1–25 (2018)
55. Estima, J., Painho, M.: Investigating the potential of OpenStreetMap for land use/land cover production: a case study for continental Portugal. In: Arsanjani, J., Zipf, A., Mooney, P., Helbich, M. (eds.) OpenStreetMap in GIScience. Springer International Publishing (2015)
56. Jonietz, D., Antonio, V., See, L., Zipf, A.: Highlighting Current Trends in Volunteered Geographic Information, pp. 1–8 (2017)

Generating 3D City Models from Open LiDAR Point Clouds: Advancing Towards Smart City Applications

Sebastián Ortega, José Miguel Santana, Jochen Wendel, Agustín Trujillo, and Syed Monjur Murshed

Abstract In the past years, the amount of available open spatial data relevant to cities throughout the world has increased exponentially. Many cities, states, and countries have provided or are currently launching the provision of free and open geodata through public data portals, web-services, and APIs that are suitable for urban and smart cities applications. Besides ready to use 3D city models, many free and open LiDAR data sets are available. Several countries provide national LiDAR datasets of varying coverage and quality as free and open data. In this research, we introduce a novel pipeline to generate standardized CityGML conform Level of Detail (LoD)-2 city models for city-wide applications by using LiDAR generated point clouds and footprint polygons available from free and open data portals. Our method identifies the buildings and rooftop surfaces inside each footprint and classifies them into one of the five rooftop categories. When multiple buildings are present inside a footprint, it is divided into their corresponding zones using a novel corner-based outline generalization algorithm, addressing the need for more precise footprints and models in geometric and semantic terms. Finally, CityGML 2.0 models are created according to the selected category. This pipeline was tested and evaluated on a point cloud dataset which represent the urban area of the Spanish city of Logroño. The results show the effectiveness of the methodology in determining the rooftop category and the accuracy of the generated CityGML models.

Keywords 3D city models · CityGML · LiDAR · Smart city

S. Ortega · J. M. Santana · A. Trujillo
CTIM, Universidad de Las Palmas de Gran Canaria, Las Palmas, Spain

J. Wendel (✉) · S. M. Murshed
European Institute for Energy Research (EIFER), Karlsruhe, Germany
e-mail: wendel@eifer.org

A. Mobasheri (ed.), *Open Source Geospatial Science for Urban Studies*, Lecture Notes in Intelligent Transportation and Infrastructure,
https://doi.org/10.1007/978-3-030-58232-6_6

1 Introduction

Research on Smart Cities has risen abruptly and rapidly in the last few years. It is also a major buzz word for any research and application of information and communication technology (ICT) applied to address and solve issues in cities and urban areas. The concept of a smart city thereby integrates ICT, and various physical devices connected to the IoT network to optimize the efficiency and workflow of city management, operations and services as well as to integrate citizens in these processes [1]. Smart city applications and technology facilitate local government and stakeholders to interact directly with both the community and city infrastructure and to plan, monitor and control processes in the city. A commonly used unit to store and aggregate data are thereby buildings, city furniture or dynamic or stationary city objects. Virtual 3D city models play therefore a major role in smart cities applications and research and have a long history in the application of different types of complex urban phenomena. In the past, most of the 3D city models, purely based on graphic and geometrical models, were developed for visualization purposes to ensure a photorealistic resembling of visual urban objects. In addition, recent developments focused on the inclusion of semantic aspects that are suitable for modelling and analysis purposes. For example, with the development and introduction of standardized data models such as CityGML, one can not only represent the graphical appearance of city models, but also include the representation of semantics, thematic properties, taxonomies and aggregations of different features (e.g. walls, roofs, etc.). Such virtual 3D city models form a basis for integrating numerous other items (e.g., building installation) and data (e.g. building energy characteristics) within a single framework at different Level of Details (LoD) [2]. This standard can be used as key inputs in the context of smart and sustainable cities services [3, 4]. Examples in an urban context can be found in many different domains such as noise simulation, telecommunication planning, real time monitoring, sensor data placement and integration, indoor navigation and building energy modelling, where virtual 3D city models play a major role. Biljecki et al. [5] provide a comprehensive state of the art review of applications for 3D city models were over 100 models and 29 use cases are defined.

Buildings contribute to approximately 40% of the total energy consumption of a city. Therefore, the detailed 3D information on the building stock will ensure more accurate urban energy analysis that is required for smart cities services [6]. This can help local governments to meet national and international climate goals. For example, through the geometric calculation based on the 3D building models and other building physical characteristics, dynamic heating and cooling demands for buildings can be estimated accurately [7, 8]. Furthermore, some analyses such as incoming solar energy radiation or techno-economic photovoltaic (PV) potential can be accurately calculated even at finer scale, i.e. roof or wall surface level [9]. They can be further aggregated to building or neighborhood level. In order to carry out such analyses, semantic 3D models of higher grade of representation are required [10].

Different requirements for 3D city models for visualization and modelling exist. The models for visualization purposes do not need to have geometrically correct surface models, but the analysis and simulation of urban energy related issues require geometrical and topological correct surfaces. Many 3D city models, especially those automatically generated from LIDAR or by photogrammetric methods lack correct topology and geometry. Biljecki et al. [11] reviewed the common problems associated with the CityGML data across different cities. They found some of the most common errors during the review and preparation of CityGML datasets. The errors originate due to wrong definition of normal vectors, wrong definition of building installation parts (such as building extensions should be defined as separate geometry object) and non-planarity of building surface [12]. Due to these problems, the smart city related energy models are not able to run on the concerning buildings.

Some other problems are more difficult to correct and in most cases require a significant amount of time to generate a rectified city model (e.g. nesting of polygons within one surface). The generation of simplified 3D city models (CityGML LoD1 and LoD2) from open sources is preferred and serves the requirements for energy analyses.

The number of cities represented in 3D city models is increasing rapidly, while the benefits are apparent and the investment costs and time required to build these models are decreasing as new automatic data collection technologies such as Light Detection and Ranging (LiDAR) are developed.

LiDAR is a technology that was vastly used in the last years for various tasks as 3D modelling, simulations or power management. LiDAR devices are active sensors which compute the distance between their laser sensor and the ground or an object based on the return time of a laser beam [13]. There are two types of LiDAR devices: waveform sensors, which gives the entire laser return, and discrete sensors, which gathers only a certain number of returns. The data collected with LiDAR scanners are mostly presented as point clouds, which include, as a minimum, the following data per point; a position, an intensity value and the laser return. Extra information is also added in some point clouds, such as timestamps or RGB values.

However, the generation of wide city models is still costly and time consuming. Due to this, this generation is making feasible the widespread applicability of urban analysis and modelling on a city wide scale only in a few cases. In recent years, there has been a push from local governments through larger programs such as INSPIRE[1] in Europe or the open data initiatives in many North American cities to provide open data sets that also include 3D building data. However, 3D models vary in data quality and there is a slow adoption of standards, which hinders the widespread use of such models.

Therefore, the main objective of this research is to propose a novel methodology to generate standardized 3D city models, on the basis of free and open LiDAR data that are suitable for analysis and modelling for smart cities approaches and not limited to any geographic region. This research addresses two current issues in the generation of 3D building models from open LiDAR data sets: (1) the need for a more precise

[1] https://inspire.ec.europa.eu/.

feature extraction of rooftops from LiDAR point clouds, and (2) the development of novel line generalization algorithms that can be used to automatically generate geometrically and semantically correct 3D models for the integration, analysis and modelling of smart cities applications and processes.

In the following section, an overview of the current practices for 3D building abstractions from LiDAR point clouds is given. Section 3 describes the required data sets while Sect. 4 focuses on the methodology and implementation strategies used to develop such a modelling pipeline. Section 5 introduces the results obtained over a LiDAR scanning of a major city using the proposed pipeline. Finally, a discussion and future lines of work are described in Sect. 6.

2 Related Work

During the last decade, many authors have worked on different methodologies to reconstruct 3D building models from LiDAR point clouds. Their approaches are usually classified into two categories: data-driven and model-driven [14]. Nevertheless, other categorizations can also be made, e.g., whether the proposal makes use of machine learning techniques or relies on prefixed mathematical models. Machine learning techniques seek to interpret the data with limited or no human input, but differ on how the model which gives such an interpretation is generated. Two main groups can be done: supervised and unsupervised techniques [15].

Supervised techniques require previous and hard to obtain labelled data to train an initial model for classification of new data. Resulting solutions are efficient in execution terms even with large sets. Decision trees, ensembles, support vector machines (SVM), neural networks and deep learning fall into this category. Examples applied to this field are the work of Zheng and Weng [16], which composes complex footprints into non-intersecting, quadrangular blocks and calculates different parameters for every block. A decision tree uses then the parameters to classify every block into one of 7 possible roof categories. In the work of Henn et al. [17], plane primitives are extracted by using RANSAC adjustment, which are then inserted as input for a SVM classifier. More recently, Zhang and Zhang [18] used a pipeline with convolutional, Deep-Q and residual recurrent neural networks for classifying and reconstruct city models from point clouds, and Castagno and Atkins [19] use the output of a convolutional neural network as input for SVM and decision tree classifiers for 8 types of roof models. Finally, Biljecki and Dehbi [20] use a Random Forest ensemble for prediction of the roof model, using inputs extracted from LoD 0-1 CityGML models instead of point clouds.

Unsupervised machine learning does not require labelled data and is more focused on extraction of patterns. It is easier to use at the expense of a higher resource consumption. Clustering, genetic and outlier detection algorithms are considered unsupervised machine learning. Examples of unsupervised learning applied to this field are the work of Pahlavani et al. [21], which relies on a genetic algorithm to determine which features better describe the building areas. Clustering techniques are applied in [22–25] to segment points and find candidate roof areas. RANSAC is used for plane primitive extraction in the work of Yang and Förstner [26] in a similar manner to the commented work of Henn et al. [17], but differing in the use of Minimum Description Length (MDL) instead of SVM. It is also used for extracting the normal of the best plane from triangle clusters [27].

Finally, there are other works that use classical methodologies instead of machine learning. Many of them convert the cloud into 2D images and apply region growing segmentation to find candidate roof areas [28–31]. Zhang et al. [32] combined LiDAR data and aerial imagery to generate a cost function that help optimize the geometric primitives that compose a roof. The work of Kada and McKinley [33] decomposes complex footprints and calculate their normals. Then, it establishes rules to discriminate the rooftop into 5 possible categories according to its major planes. Wang et al. [34] uses a voxel-based algorithm to segment buildings and extract roof points. In the work of Awrangjeb and Fraser [35], building masks are generated to extract points belonging to each building. Graph structures are also commonly used for creating roof feature hierarchies [36, 37] or detecting symmetries after footprint decomposition [38]. Yan et al. [39] separates non-ground points and presents a modified version of the snake algorithm to fit the surfaces. Outlines are extracted after the generation of triangles from points in [40, 41]. Finally, Awrangjeb [42] presents an algorithm to better extract edges and corners and thus generalize better a complex building footprint. Many other relevant studies that deal with 3D reconstruction of buildings from LiDAR point clouds are included in the survey of Wang et al. [43].

The following research gaps have been found:

- Most of the methods used for segmentation of rooftops in previous works are either dependent on a predefined number of clusters (k-means), too heavy in terms of computational cost (region growing) or they were proposed for detection of curved buildings [23]. There is still space for proposing a solution which does not require to set the number of clusters and can be used in a larger amount of cases.
- How to regularize a footprint is also an open problem. Common line simplification algorithms, as Douglas-Peucker, usually depend on a non-intuitive parameter which can remove critical points when not well set and thus the regularized line will not respect the original shape of the surface. Other algorithms such as the one of Zhang et al. [44] looks for detecting the two principal directions of the buildings, which implicitly assume the buildings always have a quadrangular shape and thus all angles are of 90°. Finally, snake-based algorithms either require knowledge about the expected shape or could result in an inaccurate footprint shape. There is still room to make improvements, such as proposing an algorithm which can generate footprints out of multiple building shapes with none or at least a short and intuitive interaction of the user.

Proposed research approach overcomes the research gaps and take advantage of both model and data driven approaches. Initially, we rely on clustering to find candidate roof surfaces, but using anisotropic filtering and agglomerative clustering techniques, unlike other clustering works. We also introduce a new corner-based algorithm to extract footprints. Moreover, we rely on MLESAC [45] to find plane primitives and apply rules to categorize the rooftops. We finally research the quality of generated CityGML data in widely used open-source software such as FZKViewer.

3 Required Data Sets

This research is based on free and open LiDAR and building footprint data for the generation of the 3D city models. LiDAR point clouds could be found in several formats (.las, .xyz) and there are various sources of this kind of data throughout the world. Cities such as New York (USA) [46] or Erfurt (Germany) [47], as well as entire countries, e.g. Spain, have query services to download open LiDAR data. For this research, open point cloud data from the Spanish Geographical Institute representing the city of Logroño has been used. The cloud, which can be found at [48], has 12,086,959 points and covers an area of 4 km^2 with a minimum density of 2 points per m^2. The bounding box of the chosen area is [42.4511, −2.4649; 42.4690, −2.4404]. This cloud was chosen because of having a wide variety of rooftops belonging to old and modern buildings in a relatively large extension, for which their corresponding footprints are available online. An image representing the selected area can be seen in Fig. 1.

Likewise, open building footprint datasets can be found in several web mapping services, such as Open Street Map (OSM) [49], as well as in many geoportals for local and national mapping agencies. For this study, a dataset containing 454 footprints from the same 4 km^2 area of the city of Logroño has been downloaded and processed using the Overpass API [50].

Fig. 1 Map showing the point cloud and the available footprints

4 Methodology

In this work, a pipeline for the generation of LoD1 and LoD2 CityGML models from open LiDAR point clouds is proposed, whose stages are introduced in Fig. 2.

As an overlook of the complete proposed pipeline, the process can be decomposed on the following stages:

- Given a LiDAR point cloud and a building footprint, the pipeline segments the point cloud, selecting the points that correspond to the building.
- The next step discriminates between wall and roof points, dismissing the former.
- The roof points are grouped in different roof surfaces, belonging to different buildings in the city model. For each one of them, a new footprint polygon is generated.
- For every roof surface, the algorithm looks for planes and extract features as well as intersections between those planes. Using that information, the plane will be categorized into one out of 5 possible categories: *flat*, *shed*, *hipped*, *pyramidal* or *complex*.
- Based on the assigned category, roof polygons will be generated. Finally, wall polygons are created for the final building model. The resultant models for all the buildings are included in the CityGML model.

These stages of the pipeline are thoroughly explained in the following subsections.

4.1 Wall Points Removal

Given all the points inside of a footprint, discriminating which points belong to vertical walls can be achieved by categorizing the horizontality of their associated

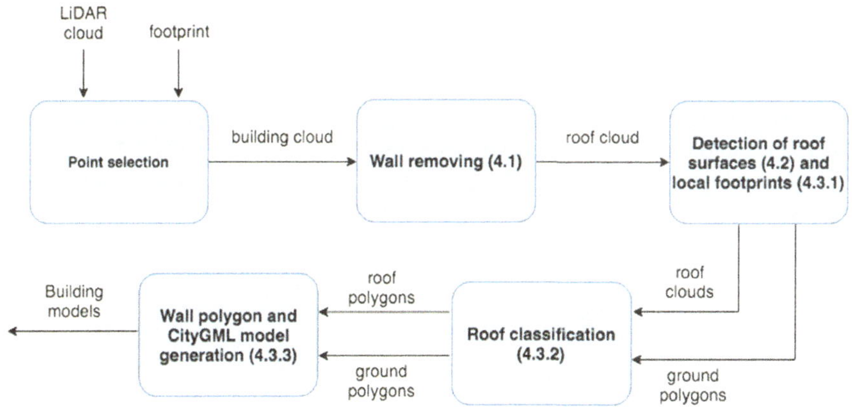

Fig. 2 Stages of the pipeline

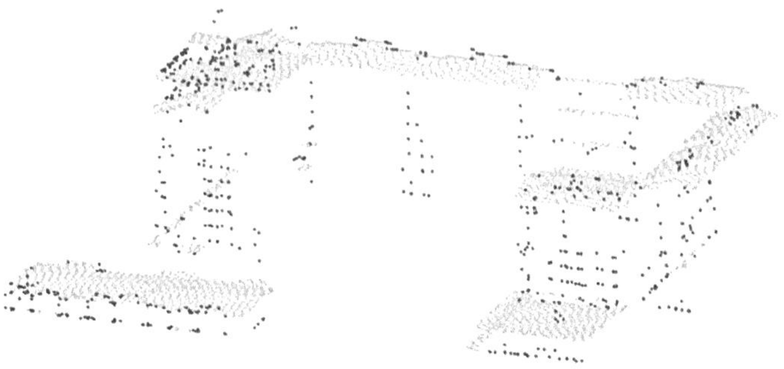

Fig. 3 Discrimination of wall and noisy points (black)

normal vectors. Our approach computes the normal vector of each point according to their 8 nearest neighbors by applying the method of Hoppe et al. [51].

After that, the verticality of the plane which best fits each point is assessed, using Expression 1, where N refers to the normal vector and Z is the upwards direction.

$$\theta_V = \arccos(|N \cdot Z|) \tag{1}$$

It is expected that a point belonging to a wall has a θ_V angle near $\pi/2$ (90°), so any θ_V in the range $[(\pi/2) - \theta_{VT}, (\pi/2) + \theta_{VT}]$ is considered to belong to a wall. Experimentally, θ_{VT} was set to 8° in this work.

The remaining roof points are then filtered to simplify the ulterior generation process. Given the normal and mean position of a point and its neighbors, a container plane is established for that wall patch. Points too distant from their corresponding planes are considered outliers due to noise or unnecessary details that will not be shown on the final model.

$$D_p = \frac{a \cdot x_0 + b \cdot x_0 + c \cdot y_0 + d}{\sqrt{(a^2 + b^2 + c^2)}} \tag{2}$$

This point-patch distance can be computed by solving Expression 2, for each point, given the plane $ax + by + cz + d = 0$ and the point (x_0, y_0, z_0). By means of D_p, the noise points are also removed from the dataset when $|D_p| > 0.2$ m.

A point will be considered *wall point* and removed if it has passed one of the two tests. In Fig. 3, it is possible to see a result of this stage.

4.2 Recognizing Different Roof Surfaces

The resulting non-wall points belong to one or multiple roof surfaces, depending on the number of adjacent buildings inside the footprint. The remaining points may also include small ground areas that should be removed. A clustering algorithm should be applied in order to segment the point cloud into different surfaces. In the choice it should be taken into account that a certain number of groups cannot be previously expected for a cloud, and also the fact that the lack of enough density in some urban clouds could make harder to detect the correct surfaces.

To overcome these issues, we apply a local transformation to the point cloud prior to the agglomerative clustering algorithm [52]. The anisotropic scaling transformation exaggerates differences in the z coordinate and decreases the differences in x and y coordinates, and it is applied to all the non-wall points as seen in Expression 3:

$$p_i^{'} = p_i \odot \left[f_{xy}, f_{xy}, f_z\right] \tag{3}$$

where f_{xy} and f_z are the factors to transform the point cloud. f_{xy} depends on the point cloud density and should always be greater than 1. f_z should be a number between 0 and 1. For this work, f_{xy} is set to 0.25 and f_z is set to 2.

Finally, we run the agglomerative clustering algorithm on the transformed cloud. Euclidean distance is used to aggregate points into a group. A cutoff, the maximum allowable distance to agglomerate points and clusters, is also defined. For this work, the cutoff is set to 1 m. An example of how different roof surfaces could be found using this stage is introduced in Fig. 4.

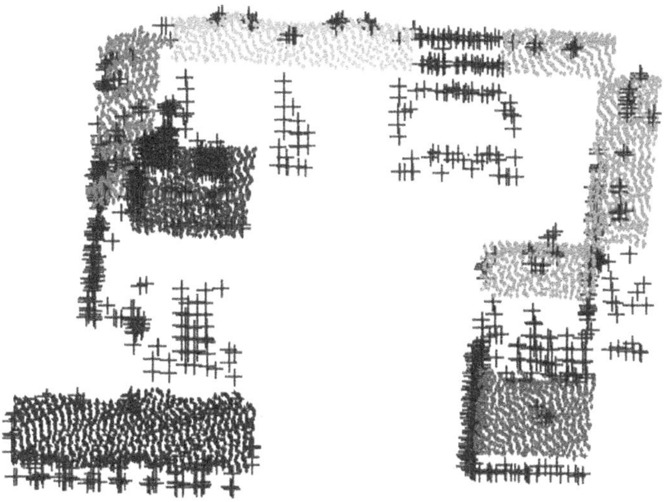

Fig. 4 Different roof surfaces found for the example of Figure B. Each one of them is shown in a different grey shade. Black crosses represent walls, ground or noise

4.3 Classifying the Roofs and Generating the Building Model

Each roof surface detected in the previous stage of the pipeline represents a building. For every building, a CityGML LoD2 model is generated. A building model requires a ground polygon, a variable number of roof polygons (depending on the type of roof) and a variable number of wall polygons. The goal is to generate a mesh of minimal complexity while still conserving the shape of the building. To perform this task, a ground polygon should first be extracted and generalized. Then, the type of roof will be estimated from the planes it consists of and their intersections, and different roof polygons will be generated according to it. Finally, wall polygons are generated between the ground polygon vertices and the roof polygon ones, as an extruded polygon with the footprint as the base.

4.3.1 The Ground Point: A Corner-Based Polygon Simplification Algorithm

To create the ground footprint, a first approximation is to generate a unique 2D α-shape [53]. The algorithm looks for the minimum largest allowed edge (the α value), which encloses the roof surface on a single polygonal boundary. The boundary created with this technique has much detail and generates small and irregular lines, that should be simplified so the representation remains close to the actual building. To do so, we applied a corner-based polygon simplification algorithm that removes needless vertices using the following strategy:

1. For each boundary point, a window of m points and the two lines before and after each point p, $ll = [p - m, \ldots, p - 1]$ and $rl = [p + 1 \ldots p + m]$, are defined. Considering the polygon as a circular list, point indices are modular. Given this, we can define the left and right difference vectors, v_l and v_r, as in the following Expressions 4 and 5:

$$v_l = \sum_{1}^{m} (ll_i - p) \tag{4}$$

$$v_r = \sum_{1}^{m} (rl_i - p) \tag{5}$$

   And an angle β_p between ll and rl is computed as in Expression 6:

$$\beta_p = \pi - \arccos\left(\left(v_l \cdot v_r'\right) \div |v_l| \div |v_r|\right) \tag{6}$$

2. The β computed for all the vertices can be expressed as a function $f(p)$ of which the local maximums are calculated. This can be appreciated in Fig. 5. All the peak points whose β value is greater than a salience threshold θ_C are considered

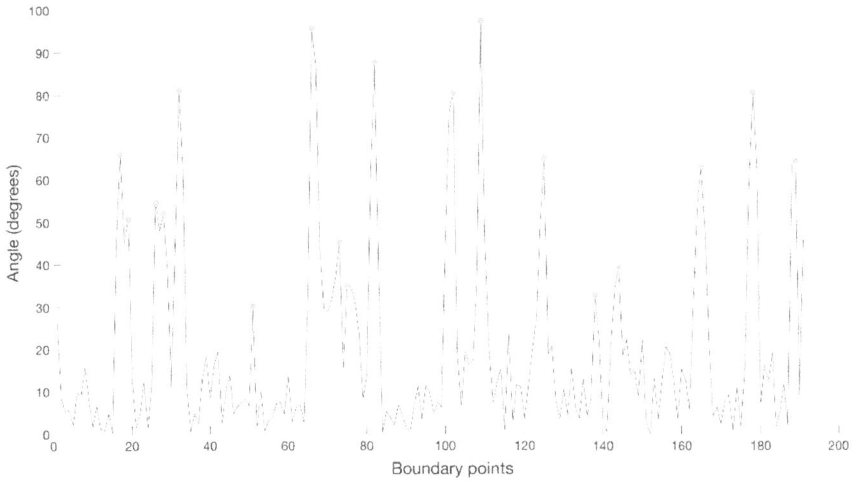

Fig. 5 Corner candidates chosen from the polygon vertices of the inverse-C-shaped rooftop from Fig. 4, based in their β values and $\theta_C = 30°$

corner candidates. θ_C can be adjusted according to the shape of the buildings in the set and the user needs.

3. For all the points between two corner candidates, including them, we fit their best line. The intersections between all the lines are then computed, and the corner positions are updated with the ones of these intersections. This results in the final footprint of the building. The final result for the example of Fig. 5 after applying this correction can be seen in Fig. 6.

4.3.2 Identifying the Category

In order to determine the category of the roof and thus create the necessary roof mesh, it is required to find all the planes within the surface. To do so, we rely on MLESAC [45], a variant of the RANSAC algorithm. The algorithm can be parametrized with an error margin to select planes, which we set on 10 cm for this work. As an example, the plane extraction done for the roof surface sample used in Figs. 4, 5 and 6 can be seen in Fig. 7.

Once the roof planes are defined, a set of several rules categorizes the rooftop:

1. There is a unique plane, or one of the planes occupies more than 70% of the rooftop area: the rooftop is classified as *flat* or *shed*. When the difference in height between all the points in the plane is below 1.5 m, the category is *flat*. In that case, the final simplified building mesh is the extruded footprint polygon with the mean z of the rooftop points as height. Otherwise, the category is *shed*. In this case, the z values for the roof polygon vertices equals the height of the closest point in the roof cloud.

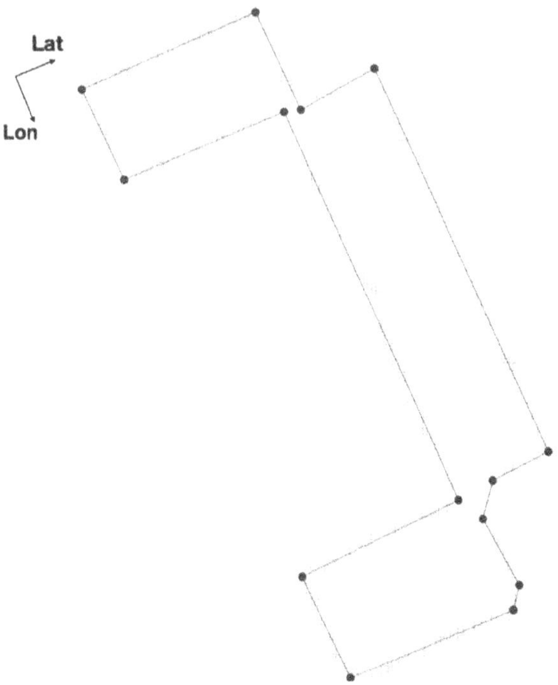

Fig. 6 Final building ground footprint for the inverse-C-shaped rooftop in Fig. 4 (grey line), against its initial boundary generated using 2D α-shape (crosses)

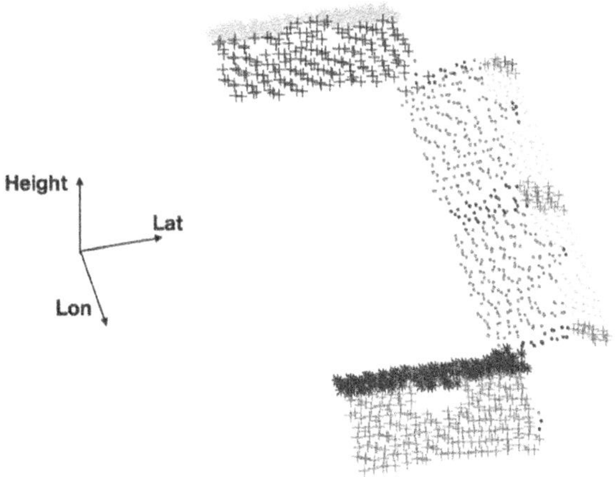

Fig. 7 Plane extraction for the complex roof Surface from Figs. 4, 5 and 6

2. There are two intersecting planes and together they occupy at least 60% of the rooftop area: the rooftop is considered hipped. In this case, two roof polygons are created in the following manner:

   a. An extruded polygon similar to the flat case is generated. The z components will be the lowest height in the roof surface.

   b. The intersection line between the two planes is calculated. From all points close to the line, both ends of the line, h_1 and h_2, are included in the polygon, between their closest two vertices. The z component of the two new vertices will be the highest in the roof surface.
   c. The final polygons are extracted from the footprint using the line vertices as pivots: $r_1 = [h_1, \ldots, h_2]$ and $r_2 = [h_2, \ldots, h_1]$.

3. There are n planes that intersect with each other and together occupy at least 90% of the rooftop area: the rooftop will be considered *pyramidal*. In this case, n roof polygons are created as a triangle fan:

   a. A base polygon is the top of the extruded polygonal footprint.
   b. The peak of the rooftop is the intersection point of all the roof planes, which can be assumed the highest point of the roof that belongs to all planes. The roof polygons consist of a pair of consecutive base vertices and the top position.

4. There are multiple planes but none of the previous rules have been fulfilled: the roof surface is categorized as *complex*. In this case, a sub-roof surface is created for any detected plane. This ensures the generation of a result, no matter the shape of the footprint: triangular or circular shapes can be processed, as well as rectangular ones. The rooftop footprint, the points inside each plane and, most importantly, the intersections between planes are considered for the generation of each sub-roof surface. When an intersection line between planes have points of the two planes near it, the polygons for both planes share vertices (corresponding to the intersection line) in order to avoid possible holes in the generated roof.

4.3.3 Wall Surface Generation

A wall surface polygon consists of a pair of consecutive ground surface vertices, g_1 and g_2, and at least two rooftop polygon vertices, t_1 and t_2, the closest to the ground surface vertices. There is a distinction between the *hipped* roof case and the general case.

In the general case, the wall polygon is generated following the order: $[g_1, t_1, t_2, g_2, g_1]$. For the special *hipped* case, two of the wall polygons include one of the line intersection points, so the polygon will be generated in the order:$[g_1, t_1, h, t_2, g_2, g_1]$. The remaining walls in the *hipped* cases will be created as in the general case.

5 Results and Discussion

Our building model pipeline was tested, as commented in Sect. 3, over a 12.1 million points, 2 points per m^2, density point cloud of the city of Logroño in Spain. Footprints of the city were collected from an OSM database. Inside the footprints, 1261 buildings

Table 1 Confusion matrix of roof surfaces

	Predicted					
		Flat	Shed	Hipped	Pyramidal	Complex
Real	Flat	372	0	4	0	1
	Shed	0	52	2	0	1
	Hipped	9	1	223	8	5
	Pyramidal	1	0	2	45	1
	Complex	5	0	14	25	490

with a minimum area of 30 m^2 can be found, from which a ground truth of roof categories was manually crafted.

A comparison between this ground truth and the pipeline results can be seen in Table 1.

A quality assessment of the roof classification per category is also performed. Four quality measurements have been explored: *recall*, also called *completeness* (rate of true positives); *precision* or *correctness* (amount of correct positives); the *F1* score, a harmonic mean of recall and precision; and the *IoU* score, a metric from computer vision which originally relates the overlapping area with the combined area between two bounding boxes, but can also be used to compare the corrected positives with the predicted and the expected ones. The formula to compute each indicator is introduced in the following expressions:

$$\text{recall} = \frac{\text{TP}}{\text{TP} + \text{FN}} \tag{7}$$

$$\text{precision} = \frac{\text{TP}}{\text{TP} + \text{FP}} \tag{8}$$

$$F1 = \frac{2}{\frac{1}{\text{recall}} + \frac{1}{\text{precision}}} \tag{9}$$

$$\text{IoU} = \frac{\text{TP}}{\text{TP} + \text{FP} + \text{FN}} \tag{10}$$

where TP stands for true positives, FP means false positives and FN refers to false negatives. The results of this assessment can be seen in Table 2.

The results offer a good accuracy of the pipeline for detection of flat and shed buildings, with both recall and precision values over 94%. For hipped class, the performance is a bit worse but still competitive, being all its quality scores around 91%. Additionally, the most correct predictions are achieved for the complex class (98.4%). However, there are still issues to be solved regarding pyramidal roof classification. Although most of the pyramidal roofs are found correctly (45/49, 91.8% of completeness), the criteria which filter pyramidal roofs are not sufficiently strict,

Table 2 Quality assessment of roof classification per class

	Flat	Shed	Hipped	Pyramidal	Complex
Recall (%)	98.7	94.5	90.7	91.8	91.8
Precision (%)	96.1	98.1	91.0	57.7	98.4
*F*1 score	0.974	0.963	0.908	0.709	0.950
IoU score	0.949	0.929	0.832	0.549	0.904

resulting in a small but appreciable number (25/534) of complex roofs misclassified as pyramidal.

In Fig. 8**a**–**e** The CityGML model generation for a sample of each roof class is shown, compared with its corresponding building point cloud.

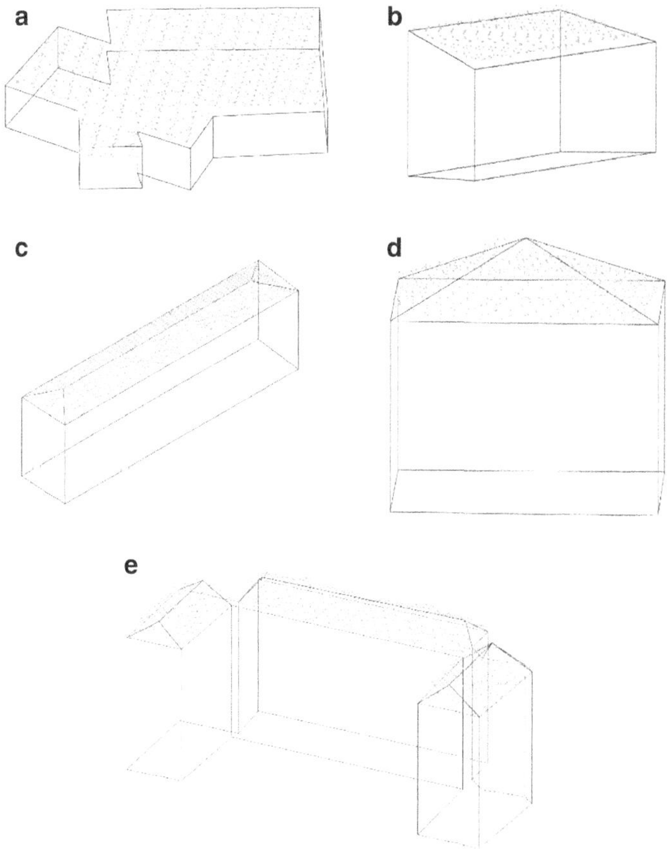

Figure 8. **a** Flat roof type, **b** Shed roof type, **c** Hipped roof type, **d** Pyramidal roof type, **e** Complex roof type (final result from example in Figs. 4, 5, 6 and 7)

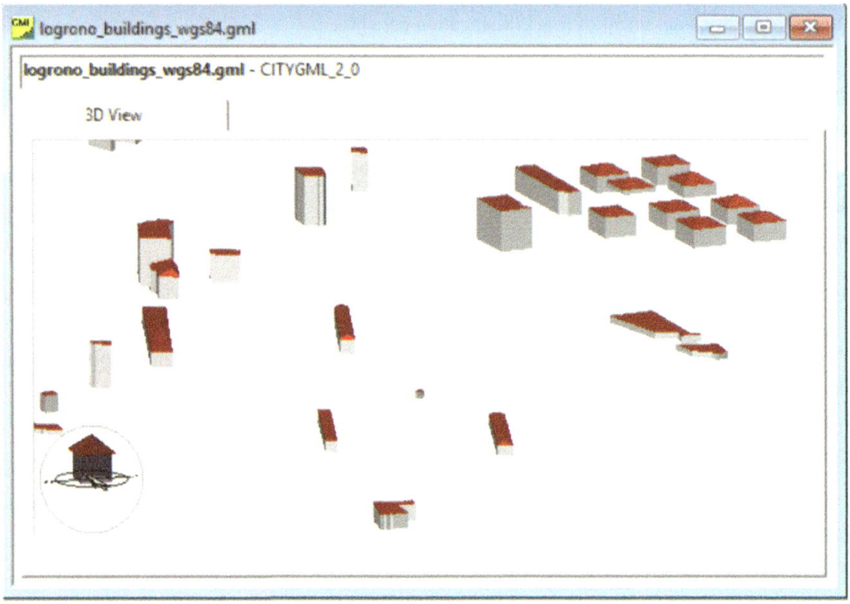

Fig. 9 Subset of buildings from the city of Logroño, generated using the proposed methodology and visualized using the FZK Viewer for CityGML models

In Fig. 9, a sample with several of the generated CityGML models for this dataset is also shown in a publicly available viewer software, demonstrating the usability of the final result.

6 Conclusions and Future Work

In this paper, we have proposed an improved methodology by combining model and data driven algorithms to extract 3D city models (e.g., CityGML LoD2) from the LiDAR point cloud. We also introduced a new corner-based algorithm to extract building footprints. Proposed methodology overcomes the limitations of current research and improves the accuracy in extracting 3D city models. It has been tested using the publicly available LiDAR datasets in the city of Logroño in Spain, covering 12.1 million points with a density of 2 points per m^2 and the footprint of 454 buildings from OSM. We observed a maximum precision of over 98% in generating shed and complex CityGML roofs/buildings, which seems promising, although a more intensive validation is still needed against benchmark sets, like the ISPRS Working Group IV one, in order to compare with other solutions. We have also tested the generation of LoD 2 CityGML, which for the current dataset translates the 245 MB point cloud into a 21 MB model. The visualization of this model has been carried out by data using FZKViewer 5.1,[2] a widely-used open source software to visualize semantic data such as BIM and 3D City Models, as well as more holistic mapping solutions, including VR and AR applications [54].

All the algorithms involved in different steps are automated in a single pipeline, meaning that considering the inputs such as a point cloud and corresponding footprints as polygons, the corresponding 3D city models are generated without user interactions. The only manual step involved in this pipeline was the identification of a ground truth for result validation.

One limitation observed in this study was the low precision (57.7%) in extracting pyramidal roofs. We did not perform analyses on point clouds originating from other sources and countries. Therefore, we would like to apply the proposed method in different sets of LiDAR point clouds in the future. Moreover, the generated LoD2 CityGML data will be evaluated against the encoding standard in order to ensure the semantical and topological correctness of the data [55]. Afterwards, an evaluation of the dataset will be made using different commercial and publicly available software such as FME workbench, FZK Viewer, 3D City database, etc.

We also intend to create a Graphical User Interface (GUI) to enable users to generate the LoD2 CityGML data from a given set of point clouds and footprints. LiDAR data are becoming very important in different smart city applications, which require large amounts of data. We have illustrated the generation of CityGML data which has direct use in applications such as building energy modelling, solar radiation or PV potential calculation. As a next step, the generated CityGML model will be used to calculate building heating and cooling energy needs as well as solar radiation and PV potential. LiDAR data will also be used to extract other three dimensional features such as road networks, electric cables or other critical infrastructures. Therefore, we strongly believe that the proposed methodology will enhance the use of LiDAR data in different emerging smart city applications.

Acknowledgements The first author wants to thank Universidad de las Palmas de Gran Canaria for its grant PIF-ULPGC-2015-ING-ARQ-2. The authors would like to acknowledge the Smart City Lab project in EIFER for partial funding of the research. Furthermore, the authors would also like to thank Alexander Simons and Alexandru Nichersu (EIFER) for their support and input in the processing and generation of the CityGML models.

References

1. Peris-Ortiz, M., Bennett, D.R., Pérez-Bustamante Yábar, D.: Sustainable Smart Cities: Creating Spaces for Technological. Springer, Social and Business Development (2016)
2. Kolbe, T.H., Gröger, G., Plümer, L.: CityGML—interoperable access to 3D city models. In: Proceedings of the First International Symposium on Geo-Information for Disaster Management. Springer, Berlin (2005)

[2]https://www.iai.kit.edu/english/1648.php.

3. Becker, T., Nagel, C., Kolbe, T.H.: Integrated 3D modeling of multi-utility networks and their interdependencies for critical infrastructure analysis. In: Advances in 3D Geo-Information Sciences, pp. 1–20. Springer, Heidelberg (2011)
4. Döllner, J., Baumann, K., Buchholz, H.: Virtual 3D city models as foundation of complex urban information spaces. In: Proceedings of Sustainable Solutions for the Information Society—11th International Conference on Urban Planning and Spatial Development for the Information Society, pp. 107–112
5. Biljecki, F., Stoter, J., Ledoux, H., Zlatanova, S.: Çöltekin, A: Applications of 3D city models: state of the art review. ISPRS In. J. Geo-Inf. **4**(4), 2842–2889 (2015)
6. Bahu, J.M., Koch, A., Kremers, E., Murshed, S.M.: Towards a 3D spatial urban energy modelling approach. Int. J. 3-D Inf. Model. (IJ3DIM), **3**, 1–16 (2014)
7. Murshed, S.M., Picard, S., Koch, A.: Modelling, validation and quantification of climate and other sensivities of building energy model on 3D city models. ISPRS Int. J. Geo-Inf. **7**, 447 (2018)
8. Agugiaro, G.: Energy planning tools and CityGML-based 3D virtual city models: experiences from Trento (Italy). Appl. Geomat. **8**, 41–56 (2016)
9. Murshed, S.M., Lindsay, A., Picard, S., Simons, A.: PLANTING: Computing high spatio-temporal resolutions of photovoltaic potential of 3D city models. In: Mansourian A., Pilesjö, P., Harrie, L., van Lammere, R. (eds.) Geospatial Technologies for All—Lecture notes in Geoinformation and Cartography. Springer International Publishing AG, Cham (2018)
10. Nouvel, R. Zirak, M., Dastageeri, H., Coors, V., Eicker, U.: Urban energy analysis based on 3D city model for national scale applications. In: IBPSA Germany Conference, 83–90 (2014)
11. Biljecki, F., Ledoux, H, Du. X., Stoter, J., Soon, K.H., Khoo, V.H.S.: The most common geometric and semantic errors in CityGML datasets. ISPRS Ann. Photogramm. Remote Sens. Spat. Inf. Sci. **4**(2):13–22 (2016)
12. Wendel, J., Simons, A., Nichersu, A., Murshed, S.M.: Rapid development of semantic 3D city models for urban energy analysis based on free and open data sources and software. In: Proceedings of the 3rd ACM SIGSPATIAL Workshop on Smart Cities and Urban Analytics, p. 15. ACM (2017)
13. Lang, M., McCarty, G., Wilen, B., Awl, J.: Light detection and ranging: new information for improved wetland mapping and monitoring. Natl. Wetlands Newslett. **32**(5), 10–13 (2010)
14. Vosselmann, G., Maas, H.G. (2010). Airborne and terrestrial laser scanning. CRC Press (2010)
15. Flach, P.: Maching Learning: The Art and Science of Algorithms that Make Sense of Data. Cambridge University Press (2012)
16. Zheng, Y., Weng, Q.: Model driven reconstruction of 3-D buildings using LiDAR data. IEEE Geosci. Remote Sens. Lett. **12**(7), 1541–1545 (2015)
17. Henn, A., Gröger, G., Stroh, V., Plümer, L.: Model driven reconstruction of roofs from sparse LiDAR point clouds. ISPRS J. Photogramm. Remote Sens. **76**, 17–29 (2013)
18. Zhang, L., Zhang, L.: Deep learning-based classification and reconstruction of residential scenes from large-scale point clouds. IEEE Trans. Geosci. Remote Sens. **56**(4), 1887–1897 (2017)
19. Castagno, J., Atkins, E.: Roof shape classification from LiDAR and satellite image data fusion using supervised learning. Sensors **18**, 3960 (2018)
20. Biljecki, F., Dehbi, Y.: Raise the roof: towards generating LoD2 models without aerial surveys using machine learning. ISPRS Annals of Photogrammetry, Remote Sensing and Spatial Information Sciences, 27–34 (2019)
21. Pahlavani, P., Amini Amirkolaee, H., Bigdeli, B.: 3D reconstruction of buildings from LiDAR data considering various types of roof structures. Int. J. Remote Sens. **38**(5), 1451–1482 (2017)
22. Sampath, A., Shan, J.: Segmentation and reconstruction of polyhedral building roofs from aerial LiDAR point clouds. IEEE Trans. Geosci. Remote Sens. **48**(3), 1554–1567 (2010)
23. Song, J., Wu, J., Jiang, Y.: Extraction and reconstruction of curved surface buildings by contour clustering using airborne LiDAR data. Optik-Int. J. Light Electron Opt. **126**(5), 513–521 (2015)
24. Cao, R., Zhang, Y., Liu, X., Zhao, Z.: 3D building roof reconstruction from airborne LiDAR point clouds: a framework based on a spatial database. Int. J. Geogr. Inf. Sci. **31**(7), 1359–1380 (2017)

25. Jung, J., Jwa, J., Sohn, G.: Implicit regularization for reconstructing 3D building rooftop models using airborne LiDAR data. Sensors **17**(3), 621 (2017)
26. Yang, M.Y., Förstner, W: Plane detection in point cloud data. In: Proceedings of the 2nd International Conference on Machine Control Guidance, vol. 1, pp. 95–104. Bonn (2010)
27. Cheng, L., Tong, L., Chen, Y., Zhang, W., Shan, J., Liu, Y., Li, M.: Integration of LiDAR data and optical multi-view images for 3D reconstruction of building roofs. Opt. Lasers Eng. **51**(4), 493–502 (2013)
28. Mahphood, A., Arefi, H.: A data driven method for flat roof building reconstruction from LiDAR point clouds. International archives of the Photogrammetry, Remote Sensing & Spatial Information Services, 42 (2017)
29. Chen, Y., Cheng, L., Li, M., Wang, J., Tong, L., Yang, K.: Multiscale grid method for detection and reconstruction of building roofs from airborne LiDAR data. IEEE J. Sel. Top. Appl. Earth Observ. Remote Sens. **7**(10), 4081–4094 (2014)
30. Wang, Y., Xu, H., Cheng, L., Li, M., Wang, Y., Xia, N., Tang, Y.: Three-dimensional reconstruction of building roofs from airborne LiDAR data based on a layer connection and smoothness strategy. Remote Sens. **8**(5), 415 (2016)
31. Xu, Y., Yao, W., Hoegner, L., Stilla, U.: Segmentation of building roofs from airborne LiDAR point clouds using robust voxel-based region growing. Remote Sens. Lett. **8**(11), 1062–1071 (2017)
32. Zhang, W., Wang, H., Chen, Y., Yan, K.: Chen, M: 3D building roof modeling by optimizing primitive parameters using constraints from LiDAR data and aerial imagery. Remote Sens. **6**, 8107–8133 (2014)
33. Kada, M., McKinley, L.: 3D building reconstruction from LiDAR based on a cell decomposition approach. International archives of Photogrammetry, Remote Sensing and Spatial Information Sciences **38**(3) (2009)
34. Wang, L., Xu, Y., Li, Y., Zhao, Y.: Voxel segmentation-based 3D building detection algorithm for airborne LiDAR data. PloS One **13**(12) (2018)
35. Awrangjeb, M., Fraser, C.S.: Automatic segmentation of raw LiDAR data for extraction of building roofs. Remote Sens. **6**(5), 3716–3751 (2014)
36. Yang, B., Huang, R.; Li, J.; Tian, M., Dai, W., Zhong, R.: Automated Reconstruction of Building LoDs from airborne LiDAR point clouds using an improved morphological scale-space. Remote Sens. **9**, 14 (2017)
37. Wu, B., Yu, B., Wu, Q., Yao, S., Zhao, F., Mao, W., Wu, J.: A graph-based approach for 3D building model reconstruction from airborne LiDAR point clouds. Remote Sens **9**(1), 92 (2017)
38. Hu, X., Fan, H., Noskov, A.: Roof model recommendation for complex buildings based on combination rules and symmetry features in footprints. Int. J. Digital Earth **11**(10), 1039–1063 (2018)
39. Yan, J., Zhang, K., Zhang, C., Chen, S.C., Narasimhan, G.: Automatic construction of 3-D building model from airborne LiDAR data through 2-D snake algorithm. IEEE Trans. Geosci. Remote Sens. **53**(1), 3–14 (2015)
40. Varghese, V., Shajahan, D.A., Nath, A.G.: Building boundary tracing and regularization from LiDAR point cloud. In: International Conference on Emerging Technological Trends (ICETT), pp. 1–6, IEEE (2016)
41. Xu, J.Z., Wan, Y.C., Yao, F.: A method of 3D Building Boundary Extraction from airborne LiDAR point cloud. In: Symposium on Photonics and Optoelectronic (SOPO), pp. 1–4 (2010)
42. Awrangjeb, M.: Using point cloud data to identify, trace and regularize the outlines of buildings. Int. J. Remote Sens. **37**(3), 551–579 (2016)
43. Wang, R., Peethambaran, J., Chen, D.: LiDAR point clouds to 3-D urban models: a review. IEEE J. Sel. Top. Appl. Earth Observ. Remote Sens. **11**(2), 606–627 (2018)
44. Zhang, K., Yan, J., Chen, S.C: Automatic construction of building footprints from airborne LiDAR data. IEEE Transactions on Geoscience and Remote Sensing, **44**(9), 2523—2533 (2006)
45. Torr, P.H.S, Zisserman, A.: MLESAC: A new robust estimator with application to estimating image geometry. Comput. Vis. Image Underst. (2000)

46. USGS CMGP Lidar: Post Sandy (New York City), web resource: https://coast.noaa.gov/htdata/lidar1_z/geoid12b/data/4920/. Last accessed 2019/11/08
47. Download offene Geodaten Thuringen, web resource: https://www.geoportal-th.de/de-de/Downloadbereiche/Download-Offene-Geodaten-Th%C3%BCringen/Download-H%C3%B6hendaten. Last accessed 2019/11/08
48. Datos de vuelo LiDAR PNOA La Rioja, web resource: https://www.iderioja.larioja.org/vct/index.php?c=46757a356c32636e3976614332586139 5a2b352b4c773d3d. Last accessed 2019/11/08
49. OSM data download portal—Geofabrik. Web resource: https://www.geofabrik.de/data/download.html. Last accessed: 2019/11/08
50. Overpass turbo API. Web resource: https://overpass-turbo.eu/. Last accessed: 2019/11/08
51. Hoppe, H., De Rose, T., Duchamp, T., McDonald, J., Stuetzle, W.: Surface reconstruction from unorganized points. In: Proceeding of SIGGRAPH, pp- 71–78 (1992).
52. Day, W.H., Edelsbrunner, H.: Efficient algorithms for agglomerative hierarchical clustering methods. J. Classif. **1**(1), 7–24 (1984)
53. Edelsbrunner, H., Kirkpatrick, D., Seidel, R.: On the shape of a set of points in the plane. IEEE Transactions Inf. Theory **29**(49), 551–559 (1983)
54. Santana, J.M., Wendel, J., Trujillo, A., Suárez, J.P., Simons, A., Koch, A.: Multimodal location based services—semantic 3D cit data as virtual and augmented reality. In: Progress in Location-Based Services 2016, pp. 329–353. Springer, Cham (2017)
55. Gröger, G. Kolbe, T.H., Nagel, C., Häfele, K.H.: City geography markup language (CityGML) encoding standard, version: 2.0.0. Open Geospatial Consortium (2012)

Open-Source Approaches for Location Coverage Modelling

Huanfa Chen and Alan T. Murray

Abstract Location cover models aim to site one or more facilities in order to provide service to demand in an efficient way. These models are based on economic and location theory. They figure prominently in public and private sector planning, management, and decision-making. Because of broad application and extension, these models have been implemented in several geographic information system-based software packages, both proprietary and open source. Among them, open-source implementations are appealing to many because of transparency and cost-effectiveness. As usage among such approaches increases, important questions of optimality and efficiency arise, and heretofore have not been investigated. In general, there is a lack of systematic review of open-source software that supports location coverage modelling. To examine the implications of open-source approaches, a comparative assessment of the functionality provided by open-source packages enabling access to location cover models is provided. This study also suggests directions for future implementation improvement.

Keywords Spatial optimisation · Location cover model · Open source · Tool comparison

1 Introduction

Facility location modelling is critically important for both public- and private-sector application, planning, and decision-making contexts. In the public sector, service facility location selection must account for not only financial costs but also social benefits and accessibility [1]. For example, local governments need to determine the

H. Chen (✉)
Centre for Advanced Spatial Analysis, University College London, London, UK
e-mail: huanfa.chen@ucl.ac.uk

A. T. Murray
Department of Geography, University of California at Santa Barbara, Santa Barbara, CA, USA
e-mail: amurray@ucsb.edu

A. Mobasheri (ed.), *Open Source Geospatial Science for Urban Studies*, Lecture Notes in Intelligent Transportation and Infrastructure,
https://doi.org/10.1007/978-3-030-58232-6_7

locations of public libraries and schools in order to maximise accessibility to/for the public. In the private sector, retail providers must consider store siting that accounts for a range of costs, including transportation to/from distribution centres, delivery and access to serve the target customer base and revenue potential.

To identify or evaluate spatial configurations associated with location-related decision making, a range of quantitative location models have been developed [2]. In most cases, such models are used in combination with geographic information systems (GIS), as they provide the capacity to create, manage, manipulate and display spatial information. This combination enables convenient and user-friendly access to location models, and makes them part of broader spatial analysis capabilities [3]. For example, commercial GIS packages, including ArcGIS [4] and TransCAD [5], explicitly structure and solve a range of location models [3]. However, location models in these commercial packages are not without limitations and technical concerns, which include computation efficiency and unknown solution quality [3].

It is not uncommon for users and/or researchers to develop custom location modelling software, and this is common in academia. A typical approach is to formulate a location coverage problem as a mathematical optimisation problem, which is then solved using general-purpose optimisation software such as GNU Linear Programming Kit (GPLK), Gurobi Optimiser and IBM ILOG CPLEX. The appeal is that optimal solutions can be identified (or if not, then a bound of solution quality is possible), the model extended in various ways, and modification of the model is possible. In general, however, developed code is proprietary and not openly available for use by others. As a result, it is difficult to reproduce reported models and results, and the reliability of custom software is largely unknown. Moreover, developing custom software for location modelling requires knowledge and expertise in computer programming, GIS and mathematical optimisation, which may be beyond the skill set of most users, analysts, planners, decision makers, etc.

Open-source software represents one of the more promising options for broader application of location modelling, especially cover approaches. Spatial analysis more generally has long embraced open-source access to methods and systems. Examples supporting various sorts of geographical analysis include GRASS GIS (https://grass.osgeo.org/), QGIS (https://qgis.org/en/site/) and R (https://www.r-project.org), but also PySAL (https://pysal.readthedocs.io/en/latest/) and GeoDa (https://geodacenter.github.io/download.html). Rey [6] discusses the evolution of PySAL and its significance as an open-source library. In terms of continued growth and importance, the fact that GeoDa now has approximately 289,000 registered users speaks to what the future holds. This is an exponential growth rate since the 391 in April of 2005. Similar trends hold for the other spatial analysis software noted above.

The advantages of open source software are obvious. First, open source guarantees transparency and replicability of applied models, methods or algorithms as there are no longer any "black boxes" that conceal the implementation [6]. Second, open source guarantees access to and control of the source code for audit and modification, as well as the ability to redistribute code with no additional costs. Third, open source software guarantees extensibility by allowing advanced users the ability to easily customise the models to fit their needs as necessary.

While open-source location cover packages are still in the early stages of development, a number of options exist for carrying out analysis and planning. Examples include PySpatialOpt, Maxcovr, and Facility Location Problems Spreadsheet Solver (FLP Solver). Significant at this stage is the lack of systematic review focused on the state of the art associated with aspects of functionality. To this end, this study seeks to evaluate open-source software packages for location cover modelling in terms of the functions provided, representation of geographic space, and computational expense. Location cover models that are readily accessible in ArcGIS software are used as comparative benchmarks.

The next section provides background for this research. This is followed by a review of location cover models relied upon in this study. Details about the functionality of existing open-source location cover models are provided. The chapter ends with discussion and concluding remarks.

2 Background

There is an extensive range of location models that support different management, planning and decision-making contexts. Location models can be classified in many ways, including coverage, median, centre, dispersion, single facility, multiple facilities, discrete, continuous, competitive, etc. Among them, location cover models have seen broad utilisation, and have been incorporated in a variety of commercial and open-source software. Since the pioneering work of Toregas et al. [7] and Church and ReVelle [8], coverage models have been used to address emergency response, cellular communication, reserve design, warning notification, mail delivery, and many others [2].

Two prominent coverage approaches have continued to attract attention, namely the location set covering problem (LSCP) and maximum covering location problem (MCLP). The LSCP [7] was proposed to represent the planning situation where the fewest facilities are to be located such that all demand is served within the designated maximum service response time or distance. In contrast, the MCLP [8] was introduced to identify a configuration of facilities fixed in number that serves the greatest amount of potential demand within the maximum service standard. The LSCP and MCLP have been examined and extended by many researchers. A comprehensive review of these models can be found in Church and Murray [2].

To support application, the LSCP and MCLP have been solved using a range of exact and heuristic approaches, and have been implemented in a variety of software packages. According to the user group (e.g., specific organisation versus mass market) and the code availability, these packages can be classified into three types:

(a) Custom software, developed for and used by a specific organisation;
(b) Closed-source software, developed for mass markets but the code is not publicly available;

(c) Open-source software, developed for mass markets with the code publicly accessible.

As reflected in (a), researchers have structured and solved location models directly [9, 10], calling general-purpose optimisation software to derive optimal location covering model solutions. While custom software allows for better control of the approach and solution quality, it comes with obvious disadvantages. Programming involves software development, dealing with data input, model implementation, coordination with solvers, etc. This is time-consuming and challenging. Moreover, in many studies, researchers generally do not reveal details of the software, implementation specifics or utilised parameters, nor do they publicly share the code. As a result, these studies are difficult to verify and reproduce, which ultimately impedes the development and application of location models.

Representative of option (b) are the Location-Allocation function in ArcGIS (accessible through Network Analyst Tools) and the Facility Location Model function in TransCAD. These commercial tools are used to make plans and/or support decision making, but what is rather remarkable is that no attention is given to solution quality and reliability issues. This is important because these commercial packages rely on heuristics for solving location models, and do not provide information on solution details or quality assessment of results. The significance of such a point can be found in Woodhouse et al. [11], who reported that ARC/INFO did not find the optimum for LSCP instances examined, and Murray et al. [3], who found that ArcGIS and TransCAD failed to identify an optimal solution in the majority of MCLP and LSCP problem instances (1059 evaluated). This is not necessarily surprising given that heuristics are adopted in these packages, but since the solution approaches involve multiple parameters that influence speed and solution quality it is an issue that there is no mechanism for informing users and/or allowing them to interact with associated parameters [12, 13]. Depending on the context, such heuristic approaches and non-optimal solutions may raise issues of significance.

Open-source software, option (c), provides an appealing alternative for using and applying location cover models. Open-source software is effectively code that anyone can inspect, apply, modify and enhance. Several open-source packages supporting location cover modelling have been implemented and released. Among them, a Python-based package PySpatialOpt [14] contains a wide range of cover options, including MCLP and LSCP. It has bindings for GIS software, including Esri ArcGIS and QGIS, to generate facility service areas or facility-to-demand distance matrices, and then it calls a linear-integer programming solver for problem instance solution. There is also Maxcovr [15], an R package tool that provides access to MCLP and LSCP. Although both packages utilise linear-integer programming and theoretically can identify optimal solutions, there are considerable differences between them in terms of model types and model assumptions. There is also FLP Solver [16] that provides solution to the MCLP and LSCP, which is based on Microsoft Excel and Visual Basic for Applications. Unlike PySpatialOpt and Maxcovr, this package utilises heuristic methods to solve the MCLP and LSCP.

It is noteworthy that open-source location cover model approaches are available to support analysis, planning, management and decision making. However, it is critical to review these packages comparatively with widely used commercial alternatives. The reason is that assessment along these lines facilitates identifying potential issues and limitations. Further, recognising opportunities for future research in open-source location cover modelling is important given more general trends of growth, development and utilisation noted in Murray et al. [3].

3 Methods

As mentioned above, two location cover models are of interest in this study, namely the LSCP and MCLP. In this section, mathematical formulations are presented to describe the models and show how they differ, as well as how they have been extended.

The following notation is considered:

i: index of demand units, $i = 1, \ldots, n$

j: index of potential facility sites, $j = 1, \ldots, m$

d_{ij}: shortest distance or travel time from demand unit i to potential facility site j

S: maximum service distance or time standard

N_i: set of potential facility sites j that can cover demand i. It can be defined as $N_i = \{j|d_{ij} \leq S\}$, accounting for whether a potential facility site j is within the service distance/time standard of a demand unit i.

$$x_j = \begin{cases} 1, & \text{if a facility is sited at location } j \\ 0, & \text{otherwise} \end{cases}$$

With this notation, the LSCP is as follows [7]:

$$\text{Minimise} \quad \sum_j x_j \tag{1}$$

$$\text{Subject to} \quad \sum_{j \in N_i} x_j \geq 1 \quad \forall i \tag{2}$$

$$x_j \in \{0, 1\} \quad \forall j \tag{3}$$

The LSCP objective, (1), is to minimise the number of facilities sited. Constraints (2) guarantee that each demand unit is covered by at least one sited facility. Constraints (3) impose binary restrictions on location decision variables.

The LSCP considers the number of facilities as the significant factor in the decision making process [8]. However, in cases where total resources are insufficient to achieve coverage of all demand, decision makers may seek to cover as much demand as possible by the number of facilities that can be afforded. This is the motivation of

MCLP, which relaxes the requirement that all demand be served within the service standard.

The following additional notation is introduced:

a_i: estimated service need for demand unit i

p: the number of facilities to be sited

$$y_i = \begin{cases} 1, & \text{if demand unit } i \text{ is covered} \\ 0, & \text{otherwise} \end{cases}$$

The MCLP detailed in Church and ReVelle [8] is as follows:

$$\text{Maximise} \quad \sum_i a_i y_i \tag{4}$$

$$\text{Subject to} \quad \sum_{j \in N_i} x_j \geq y_i \quad \forall i \tag{5}$$

$$\sum_j x_j = p \tag{6}$$

$$y_i \in \{0, 1\} \quad \forall i \tag{7}$$

$$x_j \in \{0, 1\} \quad \forall j \tag{8}$$

The MCLP objective, (4), is to maximise the total demand covered by located facilities. Constraints (5) account for coverage of a demand unit. Constraint (6) specifies that exactly p facilities are be sited. Constraints (7) and (8) enforce binary restrictions on decision variables.

Figure 1a illustrates the solution for the MCLP when p = 2 and Fig. 1b shows a solution for the LSCP. The dashed line indicates potential for service within the

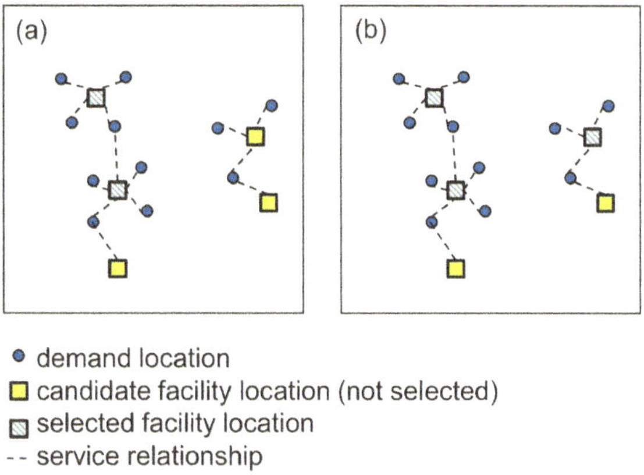

Fig. 1 An illustration of two location cover models. **a** An MCLP with $p = 2$; **b** an LSCP

service standard of a facility. Thus, in Fig. 1a two facilities are selected so as to maximise demand covered (MCLP), whereas in Fig. 1b three facilities needed in order for all demand units to be covered (LSCP).

As with any model, there are different variants and extensions possible. This is true for both the LSCP and MCLP. An overview of coverage model variants and extensions are detailed in Church and Murray [2]. The spatial nature of the LSCP and MCLP is represented by the decision variables corresponding to the location selection of a facility, x_j, but also in the set N_i as it defines the facilities that are able to serve a demand unit i within the service response standard. This standard is often represented by a maximal service distance S, which is dependent on the underlying travel space (e.g., Cartesian, network, etc.) and movement potential (e.g., geodesic, Euclidean, rectilinear, network, etc.).

The representation of demand too can mean variation in modelling approach. Demand units are spatial features and can be represented as points, lines or polygons, and the coverage provided to each demand can be classified as binary and partial. Binary coverage indicates that demand i can be either covered or not covered by each facility. In contrast, partial coverage denotes that demand unit i can be partially served by a facility, possibly depending on covered areal size [17]. Additionally, demand units may be homogeneous or heterogeneous in terms of spatial distribution. While the MCLP is defined with respect to heterogeneous demand through the use of a_i, variation too is possible. Specifically, it may be that a uniform weight ($a_i = 1$) is relied upon, which can be found in the Maxcovr package.

A final variant in modelling approach is the addition of service facility capacity constraints. Such a constraint is appropriate in applications where facilities experience limitations is services that can be provided. This means that the capacities should not be exceeded for each facility. The formulations of the capacitated LSCP and capacitated MCLP can be found in Church and Murray [2]. Noteworthy is that the incorporation of capacity constraints on facilities makes problem solution much more difficult [18].

4 Access and Solutions

A workflow of using a location cover model is summarised in Fig. 2, and each tool has assumptions and limitations on each step. In this section, a detailed comparison of the location cover model functionality provided by different packages is presented. Table 1 is a summary of the primary differences.

The discussion that follows attempts to highlight distinguishing characteristics with respect to the denoted categories given in the first column of Table 1, namely coverage computing, model type, capacity, demand unit, demand weight, space, distance metric and solution approach.

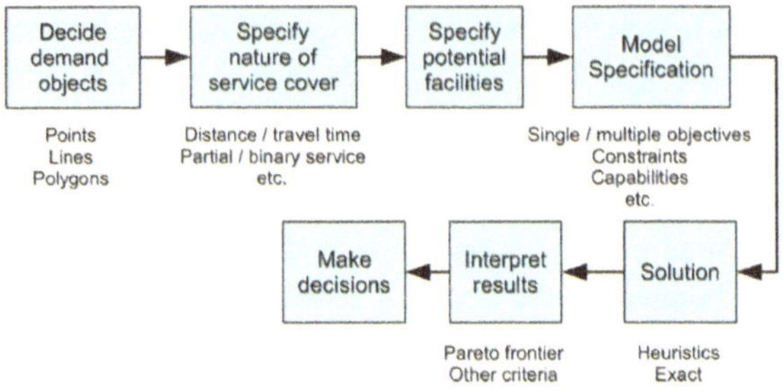

Fig. 2 Workflow of using a location cover model

Table 1 A comparison of different location cover model software

	ArcGIS	PySpatialOpt	Maxcovr	FLP solver
Coverage computing	Ad hoc	External	Ad hoc	Ad hoc and external
Model type	MCLP, LSCP	MCLP, LSCP, others	MCLP, LSCP	MCLP, LSCP, others
Facility capacity	Yes	No	No	Yes
Demand unit shape	Point	Point, polygon	Point	Point
Demand weight	Variable	Variable	Uniform	Variable
Space	Road network space	All	Longitude-latitude space	All
Distance metric	Network distance	All	Haversine distance	All
Solution approach	Heuristic	Exact method	Exact method	Heuristic

4.1 Coverage Computing

Coverage computing is a step to determine if a potential facility is capable of covering a demand, given the coordinates of demands and facilities and a service standard of distance or travel time. The coverage relationship can be represented by a service area of demands or a facility-demand distance matrix.

There are two ways to compute coverage, namely ad hoc (within the tool) and external computing (using other tools). External computing requires using another tool, but it provides users flexibility of specifying the coverage relationship. Whilst ArcGIS or Maxcovr allows only ad hoc computing, PySpatialOpt allows only external computing. In contrast, FLP Solver allows both ways.

4.2 Model Type

In ArcGIS, both MCLP and LSCP models accept point-based demand as inputs and consider only binary coverage. That means, in planning situations involving area-based demand (e.g., census blocks), the units should be further abstracted into points based on the areas' geometry (e.g., centroids) or other features (e.g., population centres) before calling the ArcGIS location models. This abstraction may have a great influence on the analysis results.

As with ArcGIS, PySpatialOpt provides access to the LSCP and MCLP. Apart from LSCP and MCLP, PySpatialOpt also provides access to the following location models [14]: threshold model; complementary coverage threshold model; backup coverage location problem; the trauma resource allocation model for ambulances and hospitals.

In terms of LSCP, ArcGIS and PySpatialOpt have a different response when 100% coverage cannot be obtained by a given facility configuration. If 100% coverage is not achievable, PySpatialOpt terminates the process without giving locating solutions. However, the LSCP model in ArcGIS will compute the maximal coverage using all facility candidates, returning the corresponding locating configuration.

Maxcovr provides access to the LSCP and MCLP. The LSCP model in Maxcovr is implemented as solving multiple MCLPs with an increasing p until the maximum coverage is obtained. Besides, the MCLP model in Maxcovr requires at least one existing facility in the problem. For problems with no existing facilities, users should generate a dummy facility that covers none of the given demand.

In FLP Solver, models provided include MCLP and LSCP with point demand and variable demand weights. This solver also provides access to the facility location problem that minimises the maximum service distance [16].

4.3 Capacity

In general, facilities may or may not be have capacities. The significance of capacities is that it limits total service allocation to any facility. For cover modelling, extensive discussion can be found in Church and Murray [2].

Facility capacities in PySpatialOpt and Maxcovr are assumed to be unlimited. That is, there is no mechanism for addressing capacity issues. For specifically addressing the LSCP and MCLP, addition of capacity for each facility is possible using ArcGIS and FLP Solver. As noted previously, the addition of capacity constraints generally increases solution difficulty.

4.4 Demand Unit Shape and Weight

Demand units are spatial features and can be represented as points, lines or polygons. This relates to how space is abstracted or aggregated. As demonstrated by Tong and Murray [17], the MCLP is sensitive to how space is abstracted (as points or polygons), and error or uncertainty is likely to be introduced by the use of a digital approximation of geographic space.

Point representation of demand units are allowed in all four packages discussed. However, representing demand as polygons is only possible using PySpatialOpt. If demands are modelled as polygons, two coverage types are provided: binary and partial. With binary coverage, a facility covers a demand unit if and only if the service area of the facility contains the demand area. With partial coverage, the facility-to-demand coverage is proportional to the intersection between the facility service area and the demand area.

In the MCLP, demands can be heterogeneous with variation in demand weights a_i, which are supported in ArcGIS, PySpatialOpt and FLP Solver. On the other hand, the Maxcovr package only allows a uniform weight ($a_i = 1$). Accommodating variable demand weights would require the generation of multiple units at one location, with the unit number equal to the original demand. This workaround generally increases problem size and computational cost.

4.5 Space and Distance Metric

As defined in the LSCP and MCLP formulations, the distance metric or travel time d_{ij} is important in location modelling as it helps to determine the coverage relationship.

ArcGIS requires a transport network to apply and solve location cover models, which means it supports only network space and distance. In fact, the requirement of a network can be considerably limited in the cases where no network is available and where the Euclidean distance is used. In order to support non-network distance in ArcGIS location cover models, a workaround is to create a network beforehand.

Maxcovr supports exclusively the latitude–longitude coordinate system and the Haversine formula distance [19], which assumes that the earth is a sphere and calculates the greater circle distance between two points from their longitude and latitude. The Haversine distance is a useful approximation for small-scale distance, with the error within metres. The Haversine distance d between two points is shown in (9):

$$d = 2r \sin^{-1} \sqrt{\sin^2\left(\frac{\Phi_2 - \Phi_1}{2}\right) + \cos(\Phi_1)\cos(\Phi_2)\sin^2\left(\frac{\lambda_2 - \lambda_1}{2}\right)} \tag{9}$$

where r is the radius of the sphere; Φ_1 and Φ_2 are the latitude of the two points; λ_1 and λ_2 are the longitude of the two points.

In contrast, PySpatialOpt and FLP Solver are much less restrictive in terms of space and distance metrics. PySpatialOpt supports various types of space and distance metrics. One input to this package is the service area configuration of potential facility sites or the facility-to-demand pairwise distance matrix, and this input is created using external tools. FLP Solver provides the following distance options: manual entry (aka user-define distance), Euclidean distance, Round Euclidean distance, Manhattan distance, bird flight distance (aka big circle distance) and Bing Maps driving distance.

4.6 Solution Approach

In ArcGIS, location cover models are solved using a heuristic method. This has been noted and discussed by many researchers. Briefly, a major benefit of using a heuristic to solve a location cover model is that it may identify a solution efficiently. However, there is no guarantee for most heuristics that they will derive good or optimal solutions. Moreover, even if good computation performance is observed for a heuristic for one location cover problem, the performance generally does not carry over when the heuristic is applied for other location cover problems. It is reported that heuristics from ArcGIS do not generate optimal solutions for even small-size problems.

The FLP Solver utilises tabu search, a heuristic algorithm, to solve a location cover model. In literature, tabu search has been used to solve the location problem of ambulance bases [20]. However, there is no guarantee for this heuristic that it will derive optimal solutions for even small-size problems.

In comparison, both PySpatialOpt and Maxcovr generate optimal solutions for a location cover model if such solutions exist. These solutions are obtained by calling optimisation software. The PySpatialOpt supports almost all mainstream optimisation software, including lp_solve [21], Gurobi, CPLEX, XPRESS, and GLPK. The Maxcovr supports several optimisation packages, which include lp_solve, Gurobi, and GLPK.

5 Conclusions

This chapter has evaluated location cover models in open-source packages as well as commercial software in terms of capabilities and solution optimality. Specifically, the LSCP (location set covering problem) and the MCLP (maximal covering location problem) were considered. Both LSCP and MCLP are accessible in ArcGIS, PySpatialOpt, Maxcovr, and FLP Solver.

In terms of problem assumptions, PySpatialOpt and FLP Solver are more flexible than the others as these two packages accommodate all types of space and distance metrics. In comparison, ArcGIS requires a transport network and the network distance, and Maxcovr supports exclusively latitude–longitude space and the Haversine distance. In PySpatialOpt and Maxcovr, location cover models are solved optimally using solver software, which is fundamentally different from the heuristics adopted in ArcGIS and FLP Solver.

While open source location cover tools are appealing to many, they are in the early stages of development and there exist many application challenges. First, the tools can be made more user-friendly and versatile. For example, currently PySpatialOpt requires a pre-processing step that relies on external tools to generate service area or facility-demand matrices, and this step can be integrated into PySpatialOpt. One potential improvement to Maxcovr is to provide a wider range of models and give control to users over the distance calculation. Second, currently the tools are standalone and are only loosely linked with GIS. They can be better integrated into well-known open source GIS libraries, such as PySAL, for general use.

In the future, more work is needed in the evaluation of tools for location cover models. On one hand, it is vital to apply these tools for multiple problem instances in order to demonstrate the strengths and limitations regarding optimality and efficiency. On the other hand, these tools should also be evaluated from the user perspective in different scenarios, which will help identify the best tool option for specific needs.

Acknowledgements The authors would like to thank Dr. Nicolas Tierney from Monash University, Ms. Rui Jiang from University College London, and Mr. Aaron Pulver for their valuable comments and suggestions.

References

1. White, A.N.: Accessibility and public facility location. Econ. Geogr. **55**, 18–35 (1979). https://doi.org/10.2307/142730
2. Church, R.L., Murray, A.: Location Covering Models: History, Applications and Advancements. Springer, New York (2018)
3. Murray, A.T., Xu, J., Wang, Z., Church, R.L.: Commercial GIS location analytics: capabilities and performance. Int. J. Geogr. Inf. Sci. **33**, 1106–1130 (2019). https://doi.org/10.1080/13658816.2019.1572898
4. ESRI: ArcGIS Desktop: Release 10.6 (2017)
5. Caliper: TransCAD, Version 6.0 (2017)
6. Rey, S.J.: Code as text: open source lessons for geospatial research and education. In: Thill, J.-C., Dragicevic, S. (eds.) GeoComputational Analysis and Modeling of Regional Systems, pp. 7–21. Springer (2018). https://doi.org/10.1007/978-3-319-59511-5_2
7. Toregas, C., Swain, R., ReVelle, C., Bergman, L.: The location of emergency service facilities. Oper. Res. **19**, 1363–1373 (1971). https://doi.org/10.1287/opre.19.6.1363
8. Church, R., ReVelle, C.: The maximal covering location problem. Pap. Reg. Sci. Assoc. **32**, 101–118 (1974). https://doi.org/10.1007/BF01942293
9. Tong, D., Murray, A., Xiao, N.: Heuristics in spatial analysis: a genetic algorithm for coverage maximization. Ann. Assoc. Am. Geogr. **99**, 698–711 (2009). https://doi.org/10.1080/00045600903120594
10. Tong, D., Wei, R.: Regional coverage maximization: alternative geographical space abstraction and modeling. Geogr. Anal. **49**, 125–142 (2017). https://doi.org/10.1111/gean.12121

11. Woodhouse, S., Lovett, A., Dolman, P., Fuller, R.: Using a GIS to select priority areas for conservation. Comput. Environ. Urban Syst. **24**, 79–93 (2000). https://doi.org/10.1016/S0198-9715(99)00046-0
12. Church, R.L.: Geographical information systems and location science (2002). https://doi.org/10.1016/S0305-0548(99)00104-5
13. Murray, A.T.: Advances in location modeling: GIS linkages and contributions. J. Geogr. Syst. **12**, 335–354 (2010). https://doi.org/10.1007/s10109-009-0105-9
14. Pulver, A.: PySpatialOpt: An Open-Source Spatial Optimization Library. https://github.com/apulverizer/pyspatialopt (2019)
15. Tierney, N.: maxcovr: A Set of Tools for Solving the Maximal Covering Location Problem. https://github.com/njtierney/maxcovr (2019)
16. Erdoğan, G.: FLP Spreadsheet Solver. https://people.bath.ac.uk/ge277/index.php/flp-spreadsheet-solver/. Accessed 19 Aug 2019
17. Tong, D., Murray, A.T.: Maximising coverage of spatial demand for service*. Pap. Reg. Sci. **88**, 85–97 (2009). https://doi.org/10.1111/j.1435-5957.2008.00168.x
18. Murray, A.T.: Maximal coverage location problem: impacts, significance, and evolution. Int. Reg. Sci. Rev. **39**, 5–27 (2016). https://doi.org/10.1177/0160017615600222
19. Sinnott, R.W.: Virtues of the haversine. Sky Telesc. **68**, 159 (1984)
20. Adenso-Díaz, B., Rodríguez, F.: A simple search heuristic for the MCLP: application to the location of ambulance bases in a rural region. Omega **25**, 181–187 (1997). https://doi.org/10.1016/S0305-0483(96)00058-8
21. Berkelaar, M., Eikland, K., Notebaert, P.: lp_solve 5.5, Open Source (Mixed-Integer) Linear Programming System. https://lpsolve.sourceforge.net/5.5/ (2004)

New Age of Crisis Management with Social Media

Ayse Giz Gulnerman, Himmet Karaman, and Anahid Basiri

Abstract Social Media (SM) Volunteered Geographic Information (VGI) is gradually being used for representing the real-time situation during emergency. This chapter presents the SM-VGI review as a new age contribution to emergency management. The study analyses a series of emergencies during the so-called coup attempt within the boundary of Istanbul on the 15th of July 2016 in terms of spatial clusters in time and textual frequencies within 24 h. The aim of the study is to gain an understanding of the usefulness of geo-referenced Social Media Data (SMD) in monitoring emergencies. Inferences exhibit that SM-VGI can rapidly provide the information in the spatiotemporal context with the proper validations, in this way it has advantages to use during emergencies. In addition, even though geo-referenced data embody the small percent of the total volume of the SMD, it would specify reliable spatial clusters for the events, monitoring with optimized-hot-spot analysis and with the word frequencies of its attributes.

Keywords Social media · Volunteered geographic information · Disaster management · Spatial data mining · Text mining

1 Introduction

A disaster is defined as an emergency condition, which turns into serious casualties when it exceeds the capacity of available resources to manage it. All activities within this management are to avoid the loss of life and money [1]. International Federation of Red Cross and Red Crescent Societies declare a disaster is "a sudden, calamitous event that seriously disrupts the functioning of a community or society and causes human, material, and economic or environmental losses that exceed the

A. G. Gulnerman (✉) · H. Karaman
Department of Geomatics Engineering, Faculty of Civil Engineering, Istanbul Technical University, 34469 Istanbul, Turkey
e-mail: gulnerman@itu.edu.tr

A. Basiri
Centre for Advanced Spatial Analysis, University College London, London WC16BT, UK

A. Mobasheri (ed.), *Open Source Geospatial Science for Urban Studies*, Lecture Notes in Intelligent Transportation and Infrastructure,
https://doi.org/10.1007/978-3-030-58232-6_8

community's or society's ability to cope using its own resources" [2]. The Emergency Event Database (EM-DAT) which was launched by the Centre for Research on the Epidemiology of Disasters (CRED) and initially supported by the World Health Organization (WHO) and the Belgian Government [3]. That global disaster event database indexed disasters by conforming at least one the following criteria; 10 or more people dead, 100 or more people affected, The declaration of a state of emergency, call for international assistance [4].

Alexander [5] classified the natural disasters with respect to many aspects like duration of impact, length of forecasting, frequency or type of occurrence etc. EM-DAT classifies disaster under two main branches as "natural" and "technological" [6]. While EM-DAT has no subcategory for a terrorist attack as disaster or emergency, [7] categorises them in a five-dimensional approach as type of disaster, duration of disaster, degree of personal impact, potential for occurrence and control over future impact. They focused on the type of disaster part on the non-natural disasters as manmade events like holocaust, kidnapping, and plane crashes that have effected mass amount of people like a coup event. Johnson [8] categorizes terrorism under the technological categories in his whitepaper. Disasters arise from hazards, which can be classified into three main categories of origin: natural, technological and environmental degradation. While natural disasters can originate from hydrometeorological, geological and biological hazards, environmental degradation is often induced by uncontrolled and unplanned human activity in nature.

This chapter focuses on a branch of technological hazard that originated from "*internal disturbances planned by a group or individual to intentionally cause disruption like riots, violent strikes, and attacks, including the act of large-scale terrorism*" as Johnson [8] mentions in his whitepaper. Houston, Hawthorne [9] also used a likely classification for the disasters as Johnson (2002) does by their causes like; natural (such as an earthquake or a hurricane), technological (such as an oil spill), or human (such as terrorism). He also take into account the consequences of the disasters while mentioning the political consequences too. Eshghi and Larson [10] have also taken into account the Canadian Disaster Database classification with five classes as biological such as an epidemic, geological, such as an earthquake, meteorological and hydrological such as a drought, human conflict such as terrorism and technological such as chemical materials. They also referred the Disaster Database Project by Green [11] where he classified the disasters as; conflict based disasters like bombing and massacre, human system failure like dam collapse and mine accident and natural disasters like earthquakes, storms etc. World Health Organization's Emergency Health Training Program for Africa classifies the disaster hazards as Natural–Physical as external like topographical or internal like tectonics and telluric, Natural–Biological as epidemics or infestations and Manmade/Technological as industrial, nuclear, chemical, fires, wars or civil strife and structural failures [12]. In addition to that, Terrorism as a disaster and an emergency issue have been assessed by Cutter [13] from the perspective of Geographical Information Science. The study draws perspective from both emergency management practitioners and first responders (such as; police, fire emergency, and medical teams) and emphasize the critical need of real-time data from the field.

All phases of Disaster Management (DM) that are preparedness, mitigation, response and recovery are crucial. However, the response phase can be seen the most crucial one due to race against time for the use of scarce resources. While all levels of the authorities have responsibilities to facilitate the management periods, communities, non-governmental organizations and individual contributors play important roles in providing information pertain to disaster effects on the field for this phase [1, 14–16].

The first requirement of Emergency Management is information about the affected area. Most of the disaster management systems and programs count on the estimations that are compiled before the hazard event occurred at the location like HAZUS [17], HAZTURK [18] and ELER [19]. Most of the estimations encounter uncertainty and miscalculation because of the deterministic or probabilistic scenario creation at the start [18, 20–22]. The rest of the disaster management systems count on visual interpretation, digital vectorization parts following the disaster and this takes more time than the DM can spare since a rapid response is the most important part of the DM, especially in the response and recovery phases [14]. In addition to that, Goodchild [23] questioning the capability of remote sensing for the detection of the emergency situations by asking "Could every emergency be sensed by satellites?". Beside the statistical and remote sensing techniques for disaster crisis management, the new era has started with the internet technology. One of the very first examples of the citizen science project on earthquake hazards is "Did You Feel It?" (DYFI) which is introduced in 1999. It is a pioneering automated method to collect macroseismic intensity data from internet users' shaking and damage reports. The system aims rapid information collection which is highly required for emergency during earthquakes [24]. However, this kind of project have limited by determined area and type of disaster.

The evaluation of technology in disaster management has started with the technological invent of Geographical Information Systems (GIS) in 1960s [25], continued with Peer-Production Volunteered Projects [26] and Citizen Science Volunteered Projects [27, 28] thanks to widespread use of internet in 2000s. Further developments were brought with SM platforms in late 2000s that marked a new epoch in disaster management [29–35]. Presently, SM has more than 2 billion active bionic sensors (users) who sense and share what is happening around them [36]. They produce real-time data from most of the world where has internet infrastructures and no censorship.

Volunteered Geographic Information (VGI) is stated as crucial for rapid information gathering from disaster area regarding time scarcity and limited resources in emergencies. Initial aims of this study is to draw the perspectives of VGI and its potential contribution to the emergency management. In respect to that, introduction section is extended with three sub sections. In the first part, differences between "volunteers", platforms used for contribution and motivations behind their willingness are discussed. Second, the study literates Social Media (SM)—VGI to demonstrate the potential use in different kind of DM projects. Third, SM platforms and their added VGI features are mentioned.

Following that, a case study is conducted with SM-VGI in order to show data accountability, availability and incidence detection capacity of the SMD over space and time. The case is related with so-called coup attempt in Turkey. On the 15th July 2016 Turkey came under attack by putschists. Official sources announced that there were 250 deaths and more than 2000 injuries during this coup attempt. The attempt has been assessed as a series of emergency events under the title of "terrorism" within types of disasters. Since all traditional media sources were silent during that time, only social media served citizens to be informed and communicate each other. The coup attempt covered the whole country but for the government and emergency response teams, it was not easy to handle or manage this disaster. This indicates that sharing items on SM is valuable during such an event for both citizens, emergency response teams to be informed in real-time.

The case section is divided into four subsections. In the first subsection, the spatial variation of SMD is tested with several sharing specifications in order to find out changing spatial data accountability. In the second sub, plenty of SMD capturing tools are introduced in terms of their different purposes. The third subsection presents case data as SMD that is the only available data for the critical first hours of the attacks. In the last part of the case study, data has been basically analysed by spatially and textually to gain more insight about data evaluation in time during that kind series of emergency conditions. The fourth main section includes the debates for validation of the data by matching news collected from the trustable newspaper or TV channel sources. Lastly, the study concluded with the general overview of potentials and weaknesses of SM geo-referenced data and further required studies to enhance the understandings of data reliability.

1.1 Volunteers of VGI

The basic questions about the SM users are; "who are they?" and "how did they emerge?" The answer is brought by VGI terminology. They are the volunteers, evolved with technology, and have several different contextual names such as; participators, volunteers, and neo-geographers [30, 37]. Contexts of the VGI and its varying termed volunteers is structured in Table 1 to provide easier understanding for the literature of volunteers. Turner [38], Parker [39] and Goodchild [40] called the idea of contributing to the map production with the existing online toolsets by untrained people as "neo-geography". And the untrained producers of this new mapping idea are called "neo-geographers" means all kinds of volunteers in VGI.

Initial examples of VGI are based on participatory urban planning with the use of Geographic Information Systems that is called the Public Participation Geographic Information System (PPGIS) [43, 44]. Volunteers involved in these kinds of studies are mostly local people (such as; citizens and employees), and semi-local people (such as; tourists or visitors) [45, 46] and they are specifically named as participators. Initially, participators embody relatively small groups of volunteers via local meetings. Later it turned into larger groups of people with online PPGIS [47] and

Table 1 VGI terminology [37–39, 41, 42]

Types of VGI	Neo-geography	Kind of volunteers	Neo-geographers	Platforms
Citizen Science VGI		Public actions participators (residents, tourists, business etc.)		SoftGIS, Zooniverse, Ushahidi, Scistarters
Peer-Production VGI		Deliberate volunteers (trained or untrained contributors)		Open Street Map
SM-VGI		Unconscious volunteers (application users)		Twitter, Facebook, Google, Instagram

even become wider citizen science projects [27, 28, 48–52] with the help of online platforms such as; Zooniverse [53], Scistarters [54] and Ushahidi [55].

Deliberate volunteers are assigned as directly as main volunteers of VGI, their focus contribution is producing base maps such as; buildings, roads, places, and parks [40, 41, 56]. This second type of untrained volunteers emerged with online base map platforms [38, 39, 41]. Open Street Map (OSM) is a well-known platforms of this type of VGI, that was deployed in 2004 [56]. It has also Humanitarian Open Street Map Team (HOT) projects for the humanitarian purposes [26].

Hecht and Shekhar [37] list the volunteers in SM as unconscious who are unaware of their location information use. This unconsciousness is caused by the "terms of service" acceptance of users by simply clicking the checkbox to declare, "I have read and signed this agreement". In SM-VGI, volunteers share diverse content (such as; emotions, memories and news) with no structured topic limitation for a specific project. However, volunteers of SM-VGI become together around a topic and protest authorities, companies and regulations [57–60].

Tulloch [61] denoted that the overlap between Citizen Science and the Peer Production of VGI relies on the location investigation by individuals. Though the potential difference between Peer Production and Citizen Science relates to the purpose of motivation for contribution that Citizen Science Projects are implemented to inform planning and policy decision makers, though Peer Production systems may have no purpose other than volunteers' pleasure. On the other hand Hall, Chipeniuk [62] indicate the disparity between these two terms in Web 2.0 geospatial applications appears more semantic than real.

Goodchild [23], emphasize the all-human population as sensors around the world. According to Goodchild [23], people provide the data that they sense without any gain. On the other hand, people do not only provide data as they are. They interpret their sensations with their inner and outer visions. Ball [47] claims that citizen science projects may include biased results because participators tend to manipulate all data that they provide for their personal benefit. Although, motivations of VGI types are different, all volunteers' data depends local knowledge which has blurred the distinction on the non-expert amateur and expert for map making [40]. The advantage of SMD may arise at this point; unconscious volunteers would not manipulate the results of the studies consciously. In this study, SM-VGI is assessed as a geo-news for crisis time.

1.2 SM-VGI Studies for DM

SM after its emergence of and rapid sprawl of internet usage has attracted many researchers' attention on that topic. Initially most of the SM studies based on the new media researches including text mining and sentiment analysis. Later, the use of Global Navigation Satellite Systems (GNSS) allowed SM platforms to be used as VGI [63, 64]. An only small percent of the SMD are geo-referenced with precise latitude and longitude coordinates [65]. However, there are geo-parsing studies to extract location information from text content with text mining methodologies. These geo-parsing techniques can be vital for the time of emergencies.

Gelernter and Mushegian [66] used a geo-parsing technique to determine the locations of tweets pertain to an earthquake. For a high-performance geo-parsing technique, Gelernter and Wu [67] studied 30,000 posts on the 2011 fire in Texas. Moreover, Gelernter and Balaji [68] built a heuristic algorithm based on Named Entity Recognition (NER) in order to identify the streets and addresses, buildings, urban spaces, toponyms, place acronyms, and abbreviations.

Leetaru, Wang [69] spot on that there is a crucial requirement to have a better understanding of the twitter geography due to the wide use of SMD during emergencies. This is well recognised by the United Nations (UN), as the UN has produced their first crisis map solely based on SMD [52]. Similarly, Power, Robinson [65] used 1.8 billion tweets to monitor hazards in a crisis coordination center using text mining and machine learning algorithms. Landwehr, Wei [70] have mentioned advantageous and contribution of SMD to mitigate effects of the natural disasters. Utani, Mizumoto [71] denoted that generally, hashtags (#) are being used to classify Twitter data to utilize a high-tech response system (such as; Person Finder", "Traffic Road Information", and "Relief Supply Matching System") in Japan.

SMD is mostly used for detecting the locations of emergency events. Sakaki, Okazaki [29] developed a real-time system to detect earthquakes and send the notifications to the registered users using SMD. MeCab analyser [72] was used to categorise sentences into sets of words in Japanese for the text analysis. Another study [73] used 480,000 posts and inferred location contextually. This means inferring the location where the post is about rather than it comes from. Language word classification tools to infer the locations in the messages for the Chinese language were also used for this study [73]. These two studies adopt specifically the local language text classifiers that are important for more reliable categorization and information extraction during a disaster.

SMD is also used for early warning and rescue purposes. An ongoing study on SMD is designed to support real-time early warning and planning systems for tsunamis risk in Indonesia [70]. The study records the historical tweeting activities to estimate the population density change over time to plan the evacuation and response. Acar and Muraki [31] have tested the crisis communication on SM platforms. The results showed that Twitter is the only platform after the Great Tohoku earthquake, which was 9.0 scale Richter earthquake that hit Japan. The research investigated victims' experiences on SM during the crisis and tried to improve their communication [31]. By considering the rapid increase in the volume of SMD and the ability of being alternative of the potentially destroyed telecommunication infrastructures, Yin, Lampert [74] have proposed a notification system to enhance the awareness of emergency using some textual clustering techniques applied to high-speed stream data during disasters and crises. Wang, Ye [33] studied the characteristics of wildfire over space and time using SMD to gain more insight about SMD usage in situational awareness and disaster management. The study concluded that people have relatively strong geographical awareness during wildfire hazards and are likely to communicate regarding the situational informs on wildfire hazards, response and to show their gratitude to firefighters [33].

SM is accepted a pioneering communication way during disasters, on the other hand there are debates about reliability of the data. Castillo, Mendoza [75] have studied the SM trend topic activities after disasters such as; tsunami alerts, missing people, road conditions. The study investigated the credibility of the information considering the information propagation and false rumour propagation using heuristic-based filters [75]. Poorazizi, Hunter [76] proposed a VGI quality metric to standardize the disaster management of five headings of positional nearness, temporal nearness, semantic similarity, cross-referencing and the credibility. Another quality evaluation was conducted by Crooks, Croitoru [77]. In this study, 125,000 posts from DYFI [24] were compared with approximately 21,000 of geo-referenced tweets and concluded that Twitter data semantic filtering is limited to few words.

SMD is also used for disaster-related mapping such as; the affected area, sentiments of victims, and transportation behaviour of victims. Rosser, Leibovici [78] have proposed a method for rapid mapping of the flood inundation extent using the geotagged photographs shared by SM users, also remote sensing and topographic map data [78]. Lin and Margolin [79] looked into the emotion sprawl after a terrorist attack by analysing the SMD. The sentiment and time series analysis are applied to 180 million geocoded tweets over a one-month period. Fear sprawl was interrogated according to locational proximity, social connection and physical connection and associated with them. Another sentiment analysis techniques were applied to 97,000 tweets for operational crises management [80]. SMD allows to analyse the behavior users during and after the natural disasters, as shown by Hara [81] study, where 3,307 users with over 130,000 total tweets and approximately 23,000 geo-tagged tweets were studied to estimate the victims behaviour while returning-home modes after the Great Eastern Japan Earthquake in 2011.

In addition to all, Houston, Hawthorne [9] searched on the typology and conceptualization of the SM usage for disaster management at a number of levels. The study lists the functions of SM regarding phases of disaster management. This study can be assessed as "document and learn what is happening in a disaster" that is an initial study for the function of "provide and receive disaster response information: identify and list ways to assist in the disaster response".

1.3 Twitter and VGI Features

There are several types of SM platforms such as micro-blogs like Twitter, discussion forums like Reddit, digital content sharing platforms like Instagram, social gaming sites like Mobage, social networking sites like Facebook [9]. As literature review broadens, SM-VGI studies mostly focus on the Twitter due to the popularity of the twitter usage. Twitter launched publicly in 2006 [63]. Statista [82] declared that it has reached 330 million monthly active users as to April 2017. There are nearly 3.3 billion internet users who are active on social networks around the world, whereas active social network penetration in Turkey as of January 2018 is 63% [83, 84]. Beside its growing popularity, Twitter added geo-referencing feature into its platform in 2009. This feature is named as Tweet Geotagging API that allows extracting the geographic coordinates from the tweets. Following that, Twitter announced their "Twitter Places" in 2010 that signify a geographic area on the venue, neighbourhood, or town scale [63].

Moffitt [63] identified that twitter has three main sources to geo-reference tweets. The first one is geographical reference in a tweet message. The second is geo-tagging tweets by the selection of the user thanks to GNSS. The third one is the account profile set as home location by the user. The first one requires geo-parsing techniques for geo-referencing that is mostly returns limited success due to several Natural Language Processing (NLP) limitations. The second one is the most accepted source of geo-referencing a tweet. The third one is directly accepting the home location for all tweets geo-referencing of the user, which is not convenient for disaster cases. Even the most usable and sufficient source of geo-referencing seems to be the second one; there are several ways of manipulating the location of a tweet via several other location-based applications and sources. In the next section, a case study is conducted to show basic disaster event mapping capacity of SMD and data accountability for tweet geo-referencing.

2 Case Study

2.1 Data Accountability

Variations of tweeting preferences affects tweets' geo-referencing and it is important to know data accountability conditions for mapping. Plenty of determined preferences is applied while tweeting from the same location to map possible spatial diversity (Table 2). The real "user location" in this test is marked on maps (Fig. 2) where is the location of Istanbul Technical University Faculty of Civil Engineering, the room L-201. Totally 17 tweets are posted via Twitter, Swarm and Instagram. Swarm and Instagram have link to twitter platform for posting the same content as tweet. Test tweets are listed in Table 2 with the id, tweet, platform, connection type and details of tweet action.

The actions listed in Table 2 returned only 10 tweets inserted into database table (Fig. 1). The first six tweets have not been inserted into this spatial database since they are without spatial component, from 2 to 6 have spatial tags but those are just place names and bring no location information (as a pair of latitude and longitude). Test 7 is tweeted by checking detailed location preference on Twitter and close to the exact user location. On the other hand, while posting on Swarm and Instagram, platforms search places nearby corresponding to GNSS and suggest several places

Table 2 Detail of Tweets test list

ID	Tweet	Platform	Connection Type	Details of Tweet action
1	Test 1	Twitter	3G	Without location
2	Test 2	Twitter	3G	With location label (Istanbul, Turkey)
3	Test 3	Twitter	3G	With location label (Sariyer, Istanbul)
4	Test 4	Twitter	3G	With location label (Resitpasa, Istanbul)
5	Test 5	Twitter	3G	With location label (Maslak, Istanbul)
6	Test 6	Twitter	3G	With location label (Emirgan, Istanbul)
7	Test 7	Twitter	3G	With detailed location corresponding GPS
8	Test 8	Swarm	3G	Insaat fakultesi selection
9	Test 9	Swarm	3G	ITU selection
10	Test 10	Swarm	3G	Sariyer
11	Test 11	Swarm	3G	Istanbul
12	Test 12	Instagram	3G	ITU Insaat
13	Test 13	Instagram	3G	Istanbul Turkey
14	Test 14	Instagram	3G	ITU
15	Test 15	Instagram	3G	Emirgan Sahil
16	Test 16	Instagram	3G	Sariyer
17	Test 17	Twitter	Wi-fi	With location label (Istanbul, Turkey)

test_id integer	twitteruser character varyin	tweet character varying (254)	lat numeric	lon numeric
7	Gulnerman	Test 7	41.10506781	29.01959214
8	Gulnerman	Test 8 (@ n9aat Fak3ltesi - @itu1773 in Sstanbul, T/rkiye) https://t.co/xYbbs400LG	41.10490236	29.01907682
9	Gulnerman	Test 9 (@ stanbul Teknik niversitesi - @itu1773 in rstanbul, Tsrkiye) https://t.co/XGR4JzhzQZ	41.10604004	29.02247441
10	Gulnerman	Test 10 (@ Sar0yer - @sariyerbelediye in Istanbul) https://t.co/ijoAY9JvGJ	41.16469859	29.05694962
11	Gulnerman	Test 11 (@ 1stanbul in Istanbul, Tirkiye) https://t.co/qkNtsnxh1J	41.00479978	28.98244781
12	Gulnerman	Test 12 @ T0 0n:aat Faksltesi https://t.co/Omc8o7voob	41.10479764	29.01899292
13	Gulnerman	Test 13 @ Istanbul, Turkey https://t.co/3CxMVXFnnW	41.0186	28.9647
14	Gulnerman	Test 14 @ stanbul Teknik niversitesi (3Tx) https://t.co/NJCtKXbcbU	41.10328138	29.02333759
15	Gulnerman	Test 15 @ Emirgan Sahil. https://t.co/PMEDPHsm1N	41.10483249	29.05657758
16	Gulnerman	Test 16 @ Sar:yer, .stanbul https://t.co/SaMMoMosCw	41.1667	29.05

Fig. 1 Test Tweets database table

to be attached with the post. These platforms also allow attaching distant locations to the users' exact position by the use of POI library. Therefore, users' have the ability to manipulate their own location. In addition, assumptions in the context of GNSS measurement might disarray data as seen in the rest of 7, 8, 12 test tweets in Fig. 2a, b.

The user location is marked with the orange triangle in Fig. 2a, b. Detail point of the disarray idea is assumption context. All the test data are sufficient to be used for a relatively small-scale project such as tourism or migration inter cities or countries. On the other hand, for some large-scale projects like building and street levels, the data is required to be questioned for this detail of mapping. Tweets can be either assessed, cleaned, relocated or corrected by using clues of content, metadata. Moreover, the platform where a tweet sent from is identifiable with the "https://t.co/…" link (Fig. 1) in the tweet (Fig. 1).

The test study shows that the actions with detailed location preference on Twitter and geotagged posts on Instagram and Swarm have captured as geo-referenced tweets. While detailed location preference directly attaches latitude and longitude with the accuracy of GNSS, the other SM sourced locations might be manipulated by the user. Although georeferenced SMD embody the small percentage of all SMD and incorporates possible spatial disarray, this study carries out that data in order to demonstrate its stronger spatial correlation to the events [34].

2.2 *Disaster Event Case*

On the 15th of July, 2016 Turkey was exposed to a case of emergency, i.e. an attempt of a military coup. Although the event spread widely in many cities, the vast majority of people first learned about this emergency from SM. In the first minutes of the attempt, the media, including news channels broadcasted nothing. However, SM users raised questions on the sounds of jet planes in the capital city of Turkey and the blockage of strait bridges in Istanbul. While the traditional media stayed silent about the coup and broadcasted ordinary programs, people on SM had already started to discuss about a coup and they drew the attention of the whole world. In this study, the SM

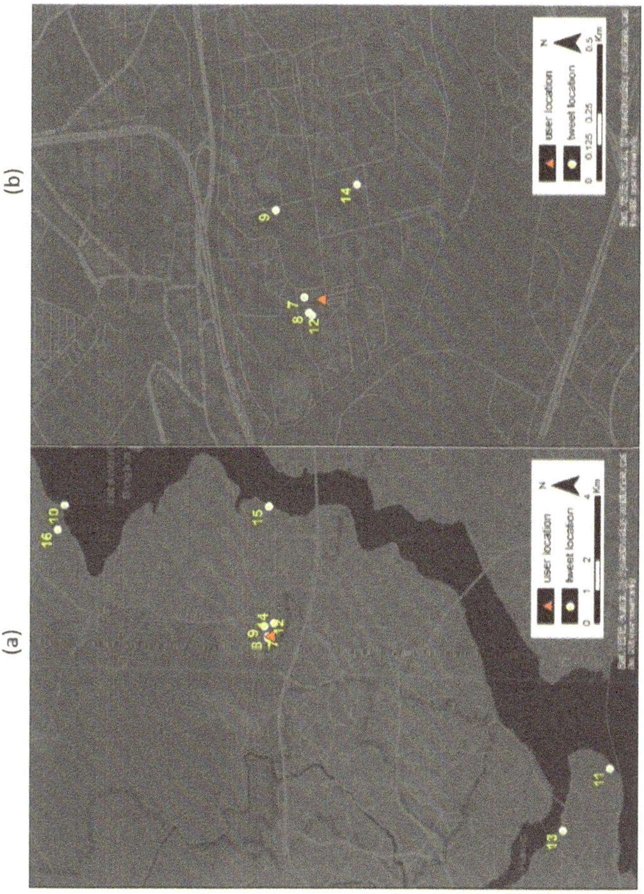

Fig. 2 Test Tweets; **a** distribution of Tweets, **b** distribution of Tweets near user exact location

reactions to a disaster were assessed by textually and spatially to show how SMD serve as a new age journalist. The following sub sections mention about this serve in the form of this disaster case in the order of; data capture tools, data overview, data process including text and spatial data analysis sub parts.

2.3 Data Capture Tools

There are diverse ways to capture and analyse SMD, a comprehensive collection of toolkits are listed by Ryerson University Social Media Lab [85]. Since the twitter data is in use of this study, prominent toolkits related with twitter data are briefly introduced. The "twitteR" [86] is an R package to download twitter data by "username", "keywords", "time interval", etc. This option allow a direct R coding environment to data analyst that can be a good start with the statistical features of R. Tweetcatcher Desktop (TCD) [87] is another free desktop program that harvests and visualizes twitter data for social sciences. Knime is a platform to provide ability of tweet harvesting with the Knime Twitter Nodes package. Knime also has SM Sentiment Analysis tools and text processes including NLP, text mining and information retrieval [88]. Carto is an online mapping platform, which provides certain amount of georeferenced SM data as free and allows drag and drop spatial analysis [89]. Moreover, there is NodeXL interface to download the data which has basic and pro versions to make text analysis and see the graphs of network between twitter users [90].

In this study, Geo Tweets Downloader (GTD) is used for data collection [91]. GTD has been designed for our previous study [42] that serves the ability of defining the spatial location, time and the text of the posts at the same time. With the help of the GTD, an analyst can collect public tweets from Twitter stream in a determined time interval based on a user-defined boundary regarding the research. GTD is a desktop application developed in Java using Twitter4j [92] library. Twitter have RESTful API [93] and StreamingAPI [94] for developers to manage, use and query Twitter functions. With the use of Twitter4j, developers can implement RESTful and StreamingAPI functions of Twitter with their own authentication parameters to their Java applications.

The application listens to stream that consist of instant tweets by twitter users who allow their tweets to be public in their privacy settings. GTD filters the tweets from the stream using Twitter RESTful API and StreamingAPI, which have geographic coordinates in a user-defined MBR (Minimum Bounding Rectangle) boundary. The application sends the filtered tweets to the PostgreSQL [95] database supported with the PostGIS [96] extension. Following this, sent data converted to a point base vector in shp (shape file) format including the attributes of "twitter_username", "tweet_text", "tweet_time" and geographic coordinates once the application detects a tweet within the user defined boundary area.

2.4 Data Overview

Case data is bounded by Istanbul that is the most populated city in Turkey. One of the first acts of the Putschists in Istanbul was the blocking of the first and second bridges over the Istanbul Strait which connect the Asia and Europe continents. This resulted in the news break since more than 360 thousand vehicles are crossing these bridges everyday. From the first tweet at 20:50 on the July 15 until the midnight of July 16, 2016, there were totally 7883 geo-referenced tweets within the determined boundary. There is no text filtering for this project data capture other than the spatial filter. Bruns and Liang [97] denotes that the first challenge of SMD capturing during crisis is finding related posts with the crisis and gathering. One approach is as used in several studies focusing on topical #hashtag contained posts which does not mean you gathered all the crises related data [97]. In this study, there is no direct focus on any #hashtags for assessing and all collected data is evaluated. Moreover, with the knowledge of geo-referenced tweets embody a small part of all tweets on the event time, all geo-referenced tweets, that are positioned either directly by GNSS or indirectly by POI libraries of other platforms, are processed. In this way, the aim of this study is constituted as to evaluate the information of the tweets, which have geographic coordinates considering both the spatial analysis and the text analysis.

Tweet counts from the start to the midnight of 16th July 2016 is displayed in Fig. 3. The number of tweets steadily decreased from 3 to 8 am. After 8 am it climbed up to 400 tweets per hour to 13:50 and afterward it slowly fluctuated until 19:50. After 19:50 it continued to rise to 600 tweets per hour.

Tweets trend for coup attempt day has been investigated with the tweets of one week before as same day and same time interval. During the coup attempt day, there are 7883 tweets tweeted by 5336 different twitter users in total while 12,033 tweeted

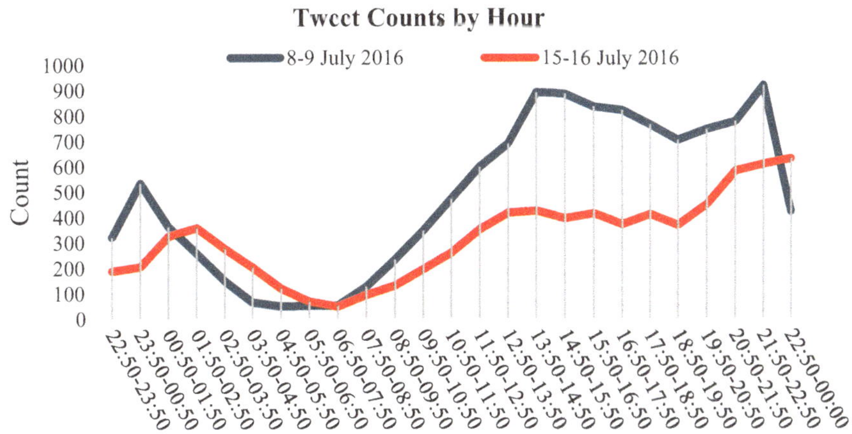

Fig. 3 Number of tweets by the hour for the event day (15–16 July 2016) and one week before (8–9 July 2016)

by 8180 different twitter users in one week before. There are 20 twitter users who tweeted more than 10 times during coup attempt and the most active user has 39 geo-referenced tweets. While one week earlier, the number of the users who tweeted more than 10 times is 12 and there is one user who has 142 tweets on that time which might cause bias and removed from the dataset. It is clearly stated that the datasets of this study do not include any bots' tweets. Bots, as automated programs, could be legitimate or malicious which generate large amount of tweets related with news and feeds or spread spams [98].

The given line graph shows the change of tweets count for coup attempt day and the same day one week before over 24 h. While the total amount of the tweets is 50% more than tweets on the day of the coup, it is the opposite between 01:50 am and 07:50 am. There is no significant anomaly for the rest of the time. This time interval covers the series of events and attacks occurred due to coup attempt.

2.5 Data Process

Data processing approach is conceptualised in Fig. 4. GTD captures only geo-referenced tweets and insert them into PostgreSQL database with PostGIS extension. Determined time interval and the spatial boundary are applied to geo-referenced data to filter. Filtered data is assessed with text and spatial analysis, which are detailed in the next subsection.

Text Analysis

The textual evaluation methodology of this study comprised text analysis to understand the SM dominance topic. The text analysis has three sub-steps; (1) pre-processing steps (Fig. 4), (2) calculating the frequency of the words, and (3) finding the correlation between words.

R programming language with RStudio is used for the text analysis [99, 100]. RStudio has both open source and commercial integrated development environment (IDE) for R that is a language and environment for statistical computing.

DBI [101], R PostgreSQL [102], and R PostGIS [103] packages were used for database connection. R Studio has text mining (TM) package [104] that enables pre-processing. The pre-processing includes the removal of white spaces, URL, and punctuations. Besides, short texts also include insignificant words that are called stop words. Since stop words has no discrimination between topics, it is meaningful to remove those words in the pre-processing part. Unfortunately, TM package serves the list of stop words for the English language but not for Turkish language, a curated Turkish stop words list [105] is scanned and removed from data.

Figure 5 presents the ordinary text cleaning steps; lowercase letters, removing stop words for English and Turkish, removing URL, removing punctuations from the corpus and stemming [106–109]. The stemming step, is to find the roots of words, that is useful for removing suffixes and counting similar words in one hand. SnowballC is a stemming package that has a Turkish extension as well [110, 111].

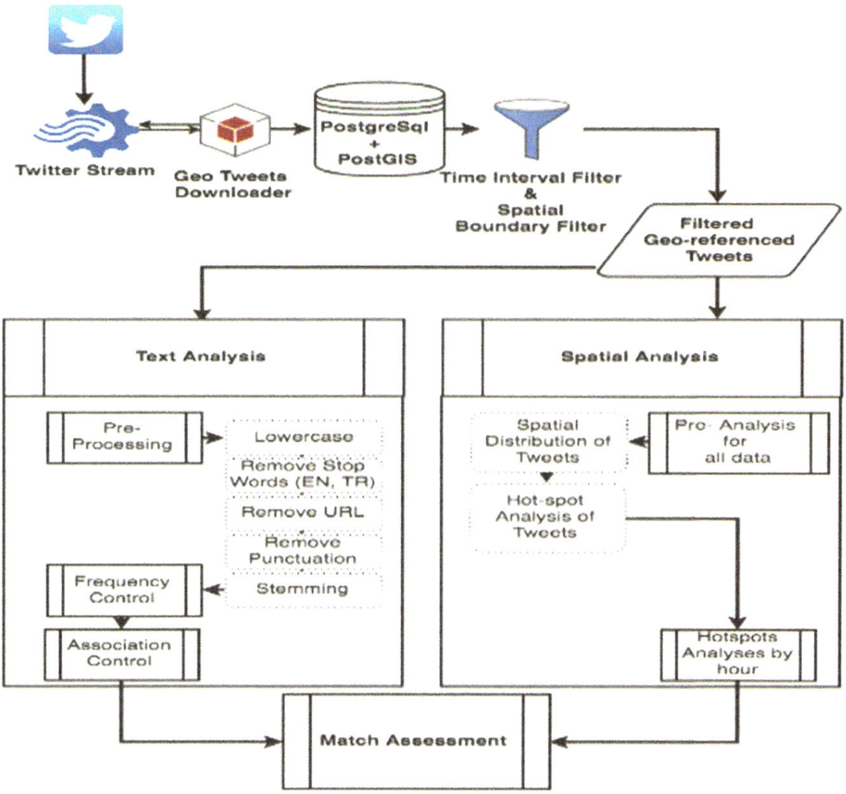

Fig. 4 Data process flowchart

Fig. 5 Pre-processing steps of text mining flow diagram

However, stemming allows us to match the words without noises of different forms of the same words, like organize, organizes and organizing [112], Turkish is not an easy language for stemming [106, 107]. Some of the words, like Turkiye (Turkey), is stemmed as "Turki", meydan (square) is stemmed as "meyda" and demokrasi (democracy) is stemmed as "demokras" in Turkish as correct as in the first form. To minimise such cases, the last pre-processing stemming step is deliberately ignored.

Following the pre-processing part, to identify the most frequently used words, document-term-matrix [113] function within TM package is used. This function categorizes words in the document into columns and the rows are filled logically considering whether or not each term in the document is used. At the end, the sum of

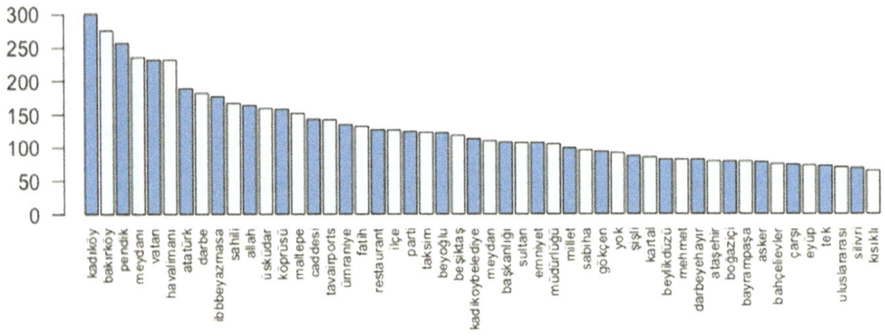

Fig. 6 The most frequent terms in the Tweet dataset within the defined time interval

the rows for each term gives the number of documents including the term. Therefore, the most frequent words are used to understand what the unconscious volunteers talked about the most within 24 h. The lack point of this technique, that would affect the result is there is no synonym or similar meaning control because the text frequency just counts the exactly the same words. The top fifty the most frequently mentioned words are shown in Fig. 6, Turkey and Istanbul, have been omitted for obvious reasons. There is a big difference between these two words and the third and the rest of the fifty words. Therefore, these words were counted as outliers to show on the bar plot and to make associations with other words.

The 50 top words list includes several district names of Istanbul and some strategic locations such as; airport names ("*atatürk*", "*sabiha*", "*gökçen*" and "*tavairports*"), bridge names ("*fatih*", "*sultan*", "*mehmet*"—"*boğaziçi*"). The rest of the words are reflecting the re related to disaster as; "*vatan*" (motherland), "*meydanı*" (square), "*darbe*" (coup), "*asker*" (soldier), "*bayrak*" (flag). Associations of these words may give us a better and systematic perspective as to what they really are instead of being estimated by an author's individual perspective. The most frequently used word was "*vatan*" (motherland), after the three districts names of Istanbul; *kadıköy*, *bakırköy* and *pendik*, which is associated more than 20% with "*bölünmez*" (indivisible), "*feda*" (sacrifice), "*bayrak*" (flag), words. The word "*meydanı*" (square) was 48% associated with "*taksim*" which indicates the heart of the city "Taksim Square". The word "*havalimanı*" (airport) and was 74% correlated with "*atatürk*", 70% with "*tavairports*", 55% with "*uluslarası*" (international), 46% with "*gökçen*", 46% with "*sabiha*", 44% with "*bakirköy*", which are all defined as the two airports of Istanbul; "*Atatürk*" and "*Sabiha Gökçen*". In addition, "*sabiha*" was 99% associated with "*gökçen*" which shows they are a completion of each other. Another word "*köprüsü*" (bridge) was 65% correlated with "*boğaziçi*", 55% with "*mehmet*", 46% with "*sultan*", 41% with "*fatih*" and specified the "*Boğaziçi*" and "*Fatih Sultan Mehmet*" bridges. Moreover, the correlations between "*mehmet*", "*sultan*" and "*fatih*" are higher than between "*köprüsü*" (bridge) which explains the completion of words. The word "*darbe*" (coup) and was 27% associated with "*girişimi*" (attempt) that clearly informed about a "coup attack". The last word for this text

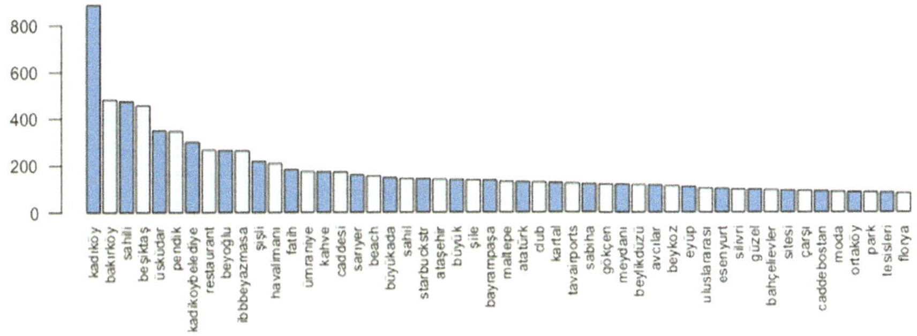

Fig. 7 The most frequent terms in the Tweet dataset one week before the defined time interval

study is "*bayrak*" (flag) which was 65% associated with "*dinmez*" (inconsolable) and 62% with "*inmez*" (not haul down), which shows the reaction of Turkish citizens living in Istanbul.

On the other hand, one week before the defined time interval, the top frequent words have been displayed in the Fig. 7. It is clearly stated, most of the districts like "*kadıköy*", "*bakırköy*", "*pendik*", "*üsküdar*", "*maltepe*", "*ümraniye*", "*fatih*", "*beyoğlu*", "*şişli*", "*kartal*", "*atasehir*", "*bayrampaşa*", "*bahçelievler*", "*eyüp*" and "*silivri*" and strategic locations names also were in the top 50 frequent words. However, the rest of the frequents words are "cafe", "restaurant", "*kahve*" (coffee), "home", "starbuckstr", "club", "park" that are mostly looks like a Friday night and weekend activities on Saturday. Moreover, there is no tweets containing "*bayrak*" (flag), "*asker*" (soldier), "*darbe*" (coup), three tweets contain "*vatan*" which are featured words within the coup day tweets dataset.

Spatial Data Analysis

Spatial analyses are to gain a general perspective of data sprawl over the defined area. They result in a spatial distribution of tweets by hour for 24 h period, the analysis of spatial distribution for one week before over the same spatial boundary. Same time interval by three hours and the comparison assessment between the coup day analyses and the analyses one week before. Spatial distribution of tweets by hour is given in Fig. 8, demonstrating the hourly change of the number of tweets and sprawl over Istanbul. As it is shown, the non-urbanized parts of Istanbul, have not much data at almost any contributed in terms of time intervals. Also, the changing spatial sprawl of data, could indicate two things: (1) there would be the emergency situations at the same location of the tweets or (2) the reactions of volunteers locating different places. In addition, the repetitiveness and intensity of the data sprawl. They seem interesting may be associated with the rise of new breaking events or the reduction of event effects. In fact, on the July 15th the crises were composed of the numerous attacks at different location and time. Some of them were instantaneous or short-lived events, like low altitude flights of jets making sonic bombs and the conflict between police and soldiers, and non-instantaneous events, like the closure of strategic points of transportation and long-term television broadcast interruption.

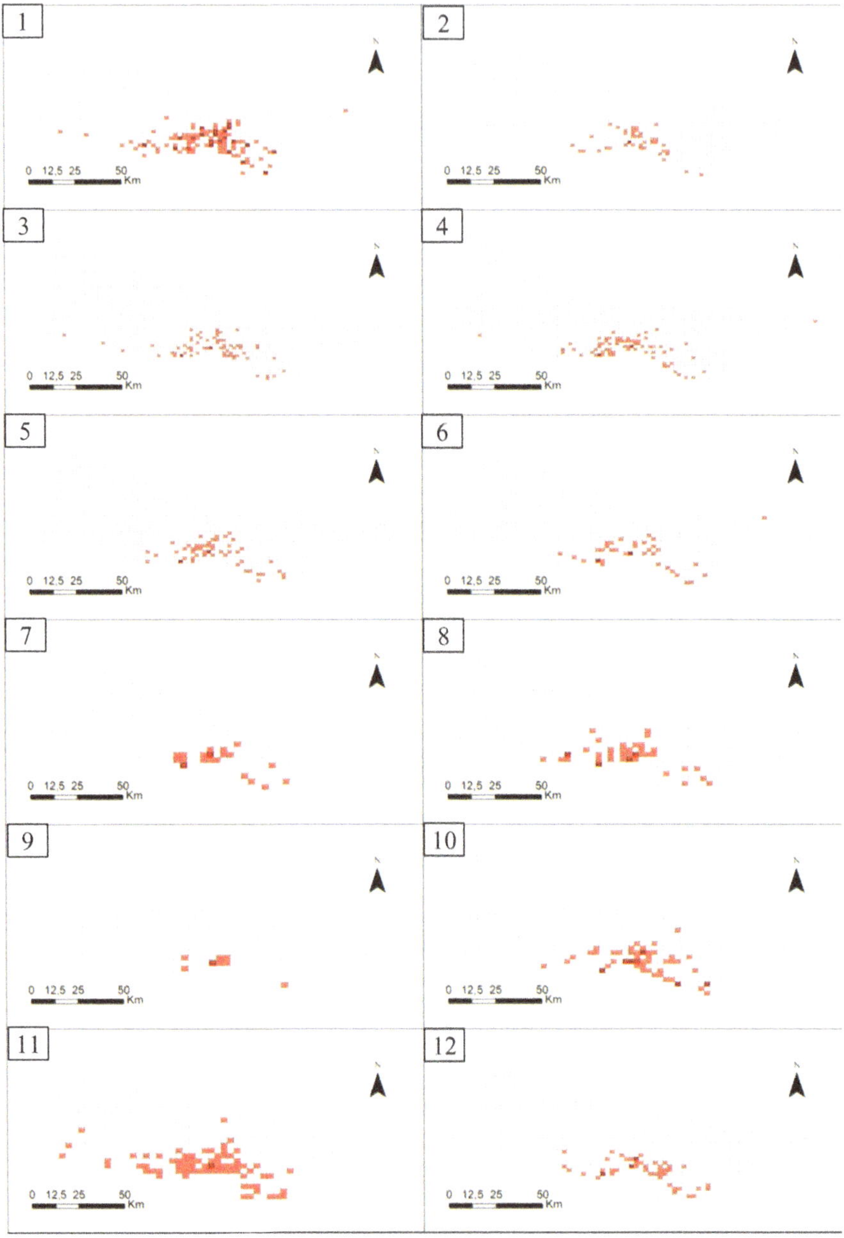

Fig. 8 Hotspots of tweets—spatial distribution for 24 h

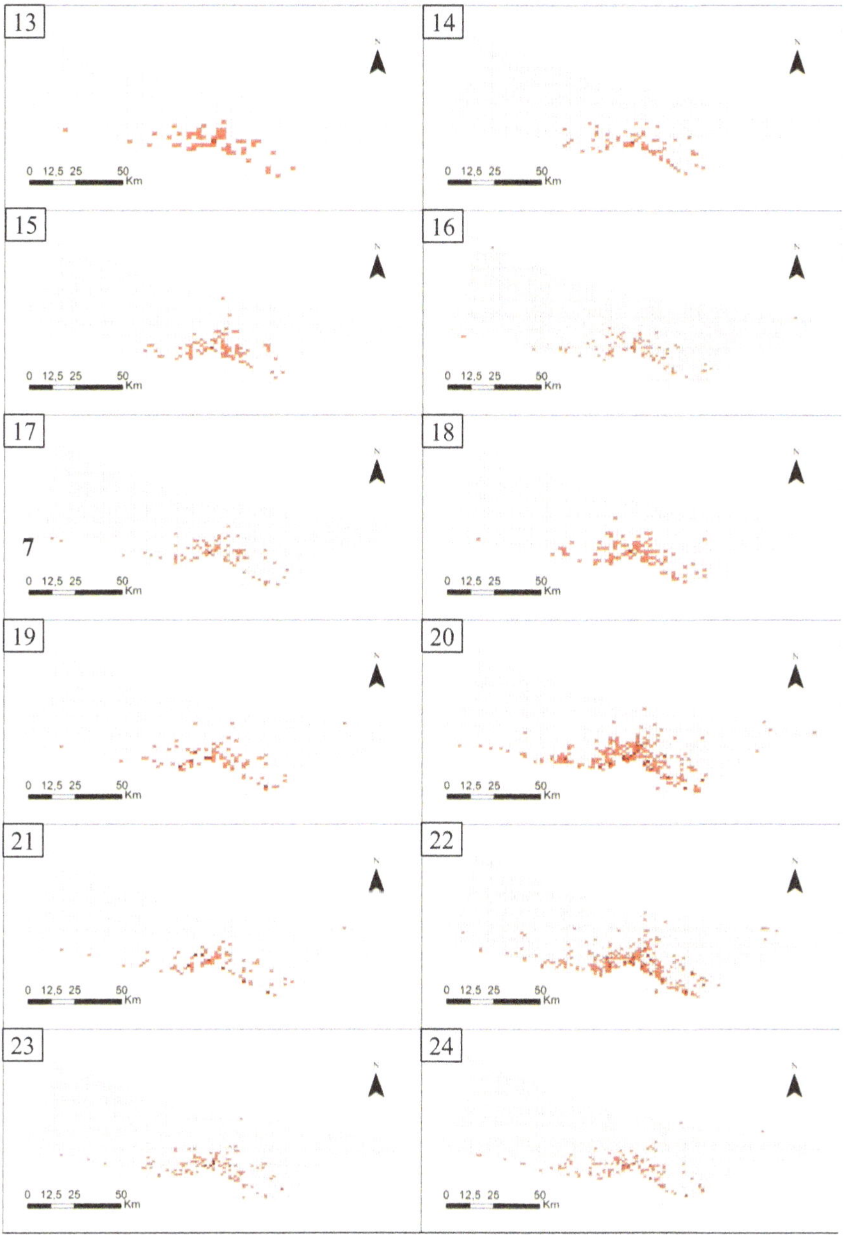

Fig. 8 (continued)

All these extractions explained above are conceptual, however, to identify more precise information extraction from spatial data, hot spot analysis was tested over hourly spatial datasets. Optimized Hot Spot Analysis [114] was used within ArcGIS software, which used the Getis-Ord Gi* statistical technique. This technique unlike the Moran I index and Geary C ratio techniques, provide with a presentation of the places of clusters [115]. Within the optimized hot spot analysis, the incident data aggregation method is chosen for "count the incidents within fishnet polygons" for defining the number of incidents within each polygon. In addition, the bounding polygons are defined as the Istanbul city boundary to compare the results of each cell, even if the cells had no incidents for each time interval. Furthermore, there was no analysis field selection to process attribute data, since all the data weights are assumed to be equal [114, 116].

The aim of the Getis-Ord Gi* statistical technique is to differentiate the clustering pixel size based on the data sprawl and density. Although the tool Optimized Hot Spot Analysis allows to analyze pixel size manually, making all pixel size equal for all-time interval may provide a lower clustering capability and potentially lower clustering accuracy. Therefore, it is better to choose Getis-Ord Gi* technique with its defining capacity of pixel size for each time interval dataset.

Optimized hot spot analysis results provide four attribute columns, which are the value of the confidence interval of each cell (z), the rest part of the confidence interval (p), the value for a cold spot, insignificance and hot spots (Gi-bin) and the number of points for each pixel (join count). Hot spots analysis of each time interval was classified into three classes, corresponding to their own natural break intervals. The darkest red pixels showed in Fig. 8 correspond the areas with the majority of the tweets as clusters that may have been assessed with the emergency.

Hourly hotspots of the Istanbul boundary provided a more detailed perception about which areas were denser, seen in Fig. 8. The size of pixels varied between the hourly analyses and provided two pieces of information. If there was a more dispersed data cluster, there are the smaller sized pixels that offered a more precise location of events. Otherwise, a bigger pixel size may indicate superclusters or less information enriched sites. So, the total number of tweets count should be considered while interpreting the data. In the context of the views above, the zero-time interval, darkest red areas, had dominancy compared to its consecutive, as interpreted as the first perception of systematic attacks at many strategic points of the city. Within the zero-time interval, Bosphorus Bridge and Fatih Sultan Mehmet (FSM) Bridge had been closed down by soldiers, Ataturk Airport and Sabiha Gokcen Airport were occupied by soldiers and air traffic was stopped and more jets started to fly at the low altitude and tanks and armoured personnel carriers had also been observed. All these unusual events attracted SM attention in the zero-time interval, as seen in Fig. 8.

All of the later time intervals had different hotspot areas, distributions and the number of hotspots pixels can vary. In some of the time interval analysis, there were many hotspots spread within the spatial boundary, like 1st, 20, 22, while some of them such as in 7th and 9th time intervals are more focused on one small or precise location. In some of the time interval analyses, there was also the dominance of interim spots with its hotspots, especially in the 1st, 3rd, 4th, and from 14 to 24th.

However, in the 11th time interval analysis, there was just one hotspot pixel with many interim pixels.

The size of pixels can vary based on the count and the spatial distribution of data (Fig. 8). The reason behind that could be the increase in the number of events in different locations. While analysing the hourly changes of an area, the trend of the actual area should be considered to avoid having a biased judgement. The area may have dominant events for the similar day or night time in the similar spatial area. In this regard, the data of an earlier week for the same spatial boundary have been studied.

Human activities on SM platform have their own spatial and temporal patterns within normal days. For example, the number of tweets has its own patterns at different hours of days of the week. The analyses on MapCalc software [117], showed how the temporal pattern during the disaster differ from the daily pattern, and displayed the spatial distribution of people during the disaster differ from normal days.

According to the intelligence reports on the newspapers beginning of the event was taken as the first-time interval, the correlation coefficient is determined as 0.695 almost highly correlated with the ordinary days' same time interval, during the expansion of the events, when most of the people are not aware of what was going on, which is second time interval, the correlation is still high according to the correlation coefficient which is 0.716. However, when the case event reach to the highest action time of the attacks where the time intervals are 3, 4, and 5, the analyses show that the correlation coefficients are low enough to prove the tweet pattern of the people is changed drastically. The correlation coefficients are determined as, −0.016, −0.003, and 0.004 for the third, fourth, and fifth time intervals, showing the tweet patterns are not the same with the ordinary times.

While results of correlation between a week before the attack and attack day dataset for the first-time interval is 0.695, the correlation is loosening as it goes further, e.g. for the fifth-time interval it is 0.004. The fifth-time interval coincides between 02:50 am and 03:50 am. This could be due to sleeping time as normally majority of people are sleeping in these times instead of tweeting or reacting to attacks. It is clearly stated that the results of correlation between datasets one week before and the attack day is decreasing considering the time-interval because attacks have been last long actively until morning. This is also interesting to justify why we need to capture and analyse the normal days data as it may give a better understanding of important intervals (e.g. sleeping period).

Moreover, the other outcome of analysis result; coincidence table explains the details of difference. The table displays comparison between value of cell through number of cells, number of cross-cell, % of cells' area, average %, standard difference and weighted score. According to the table, standard difference clarifies the change between two maps and in the hot spot values proves the change with standard difference varying between 1.97 and 2.0 except for the highest area cover group that reflects general trend in the spatial boundary.

3 Validation and Assessment

The last part of the study is about the comparison between what SMD reported and what the actual events happening. This is particularly important for credibility and reliability analysis of using SM for event detection. The events occurred for during 24-h time interval were searched in the traditional media as published. Our traditional media sources were websites of newspapers, online sources, TV channels and independent news web pages. The SMD validation for this study is done by comparing the results of text analysis and spatial analysis with the news published by trustworthy and credible such as National Newspapers and their official websites. The two instances for our media source in Fig. 9, are a newspaper named "*Hürriyet*" [118] and a TV channel named "*NTV*" [119] websites which are both well-known, old and rooted media sources in Turkey. The study has only filtered the events that have time and location information like geotags as illustrated in Fig. 9a or georeferenced and labelled on the map as shown in Fig. 9b. There were 39 events, which occurred up to 11 h after the start time of the coup attack [120].

Considering those 39 events temporally and spatially defined in the traditional media sources within our past research [120], the location of events have been summarized and sorted according to counts in Table 3. Considering that, during a 12-h time interval, the SM reaction was associated with the locations of the hotspots; however, the data that we used in our study was for a 24-h time interval, which covered a longer period. As well as this, the results of text analysis correlated with the event locations, which were extracted from traditional media [120].

(a) original news translated from Turkish

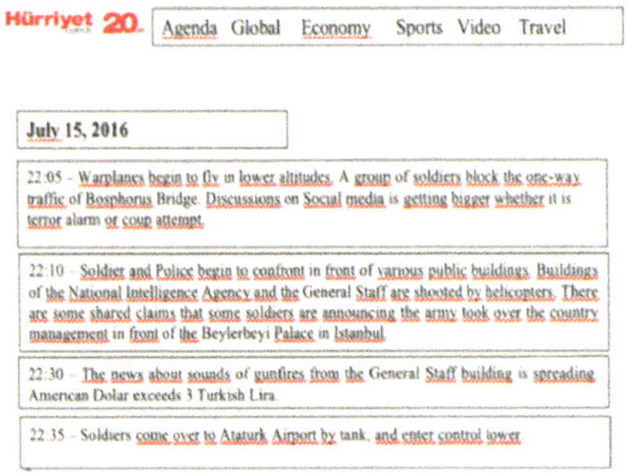

(b) original news translated from Turkish

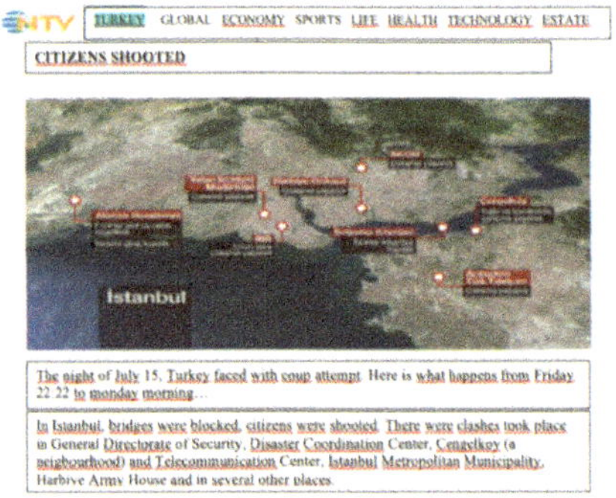

Fig. 9 Traditional Media News in online; **a** News in a newspaper website and **b** News in a TV channel website [118, 119]

Table 3 Event locations and counts in the July 15 coup attempt [120]

Event location	Event count
Bosphorus Bridge	7
Ataturk Airport	6
Istanbul (related the whole city)	5
Harbiye	4
Fatih Sultan Mehmet Bridge	3
Bagcilar	3
Sabiha Gokcen Airport	2
Taksim	2
Istanbul Strait	2
Provincial Directorate of Security	1
Bayrampasa Action Force	1
Fatih	1
Kadikoy	1
Maltepe	1
Uskudar	1
Total	39

4 Conclusion

Data sources for disaster management, especially in a time of emergency, have various channels. Expanding views of emergency mapping have enhanced the capacity building of resilience by the support of the "United Nations Platform for Space-based Information for Disaster Management and Emergency Response" (UN-SPIDER) program, since 2006. The main aim of the organization is to provide all types of space-based information usage for the whole disaster management cycle [121]. On the other hand, since 2012, GMES, GIO-EMS services in Europe have tried to support emergency management by earth observation satellites. Although, these initiatives have helped to the management of several environmental disasters, the question is "Could they serve every disaster?" The question is not just limited to the restriction of both vectorization and the attributes of event locations and conditions, but also is restricted to the question "Could every emergency be sensed by satellites?" [23]. There are disaster events like terror attacks in our case that cannot be sensed by remote sensing satellites.

Time management and the potential of response capacity building during the crisis are more important than anything, especially with scarce emergency resources, security forces, etc. The volume of SMD and count of active users have been increasing day-by-day [36] and this has already started to serve as data source for emergency management [65, 73, 74] from several different aspects and has the potential to serve much more than implemented studies [9]. While the data volume has been increasing, the process of extracting meaningful, precise and reliable information from the data is getting harder. Extracting information from this unstructured raw data stream requires pre-processing steps to clean and to test the reliability of the data, and processing steps to cluster and classify the data [33]. All SMD require a validation and accuracy test because of the unknown source of data that is provided by unconscious volunteers. However, Leetaru, Wang [69] has noted that authorities are using Twitter during major disasters to provide real-time information and directives to disaster victims while the emergency responders are initially trying to monitor Twitter as a parallel 911 emergency call system. That monitoring especially required where the telecommunication services are unavailable [74, 122]. Because of that, our further study will be on spatial reliability and textual reliability to have understanding of which data requires validation or directly used.

The most crucial part of using SM for emergency situations is monitoring by geo-referenced posts [69] which enabled around 2010 both for Twitter and Facebook [63, 64]. However, SM geo-referenced data embodies the small amount of all data [34, 65], there are several studies to enlarge geo-referenced data volume by geo parsing techniques [66, 68, 123] and to demonstrate the reliability of the geo-referenced data comparing the non-geotagged data [34]. In this study, datasets are composed of just geo-referenced data and none of geo-parsing methodologies have been applied considering geo-referenced usefulness. However, to extend the data volume in further studies, geo-parsing techniques will be considered.

VGI is seen an inexpensive, wide and sometimes quick way of data gathering; however the VGI quality is a debate regarding several conceptual quality assessments such as accuracy, granularity and consistency [124]. While researchers are studying as a peer on it, the others are claiming the data has full of unreliability [125, 126]. The reliability discussions and data cleaning perspectives are varying for SM-VGI. Some of them are trying to eliminate the reliability according to discriminate the users by using profiling [127], some of them are categorizing the accounts as human, bot and cyborg [98]. Although, within geo-referenced SMD studies includes several outlier removals in the spatiotemporal context [128–131], there is no such a study directly focusing on the ways of cleaning and testing the reliability of the spatial data gathered from SM. Within our study, there is some cleaning actions and filtering have been implemented within pre-processing steps, however deeper cleaning and filtering techniques would be used for our future studies.

In this study, the VGI concept overviewed with its place in the literature and the sub three title of VGI are explained. Moreover, as a sub title of VGI, the contribution of SM-VGI to disaster management mentioned and exemplified with various studies. In addition to that, a case study carried out as an example of series of emergency event. And the SM reaction against those emergency situations analysed textually and spatially. There are some inferences from the case. The one is that simply processes without deep analysis techniques of topic modelling and spatial statistics, SMD would provide valuable information to monitor events. The second, SMD needs to be assessed corresponding to number, duration, repetitiveness and intensity of events for better understanding of spatial distribution of reactions in time. The third, SMD requires proper validation and reliability techniques to contribute emergency management. Further studies would include the events sequential logic and the reliability of the data considering cause and result relation.

Acknowledgements This work is supported by the Scientific and Technological Research Council of Turkey (TUBITAK-2214/A Grant Program), grant number 1059B141600822, and Istanbul Technical University Scientific Research Projects Funding Program (ITUBAP-40569), grant number MDK-2017-40569.

References

1. Witt JL (ed) FEMA, partnerships in preparedness—a compendium of exemplary practices in emergency management (2000)
2. International Federation of Red Cross and Red Crescent Societies: What is a disaster? [cited 2017 02.11.2017]. Available from: https://www.ifrc.org/en/what-we-do/disaster-management/about-disasters/what-is-a-disaster/
3. CRED: EM-DAT (2016). https://www.emdat.be/
4. EM-DAT: The International Disaster Database: Frequently asked questions (2009) [cited 2017 30.10.2017]. Available from: https://www.emdat.be/frequently-asked-questions
5. Alexander, D.: Natural Disasters. Chapman & Hall. Inc., 632 pp (1993)
6. EM-DAT: General Classification (2009) [cited 2017 30.10.2017]. Available from: https://www.emdat.be/classification
7. Berren, M.R., Beigel, A., Ghertner, S.: A typology for the classification of disasters. Commun. Ment. Health J. **16**(2), 103–111 (1980)
8. Johnson R (2000) GIS technology for disasters and emergency management. An ESRI white paper (2000)
9. Houston, J.B., et al.: Social media and disasters: a functional framework for social media use in disaster planning, response, and research. Disasters **39**(1), 1–22 (2015)
10. Eshghi, K., Larson, R.C.: Disasters: lessons from the past 105 years. Disaster Prevent. Manag. Int. J. **17**(1), 62–82 (2008)
11. Green WG: Changes to the disaster database project (2003)
12. WHO and EHA: Emergency Health Training Programme for Africa (1999)
13. Cutter, S.L.: GI science, disasters, and emergency management. Trans. GIS **7**(4), 439–446 (2003)
14. Karaman, H., Erden, T.: Net earthquake hazard and elements at risk (NEaR) map creation for city of Istanbul via spatial multi-criteria decision analysis. Nat. Hazards **73**(2), 685–709 (2014)
15. Gulnerman, A.G., Goksel, C., Tezer, A.: Disaster capacity building with a GIS tool of public participation. Fresenius Environ. Bull. **26**(1), 237–243 (2017)
16. FEMA: Disaster planning is up to you (2007). https://www.fema.gov/news-release/2007/03/30/disaster-planning-you.
17. Schneider, P.J., Schauer, B.A.: HAZUS—its development and its future. Nat. Hazards Rev. **7**(2), 40–44 (2006)
18. HAZTURK: Developing a loss estimation program, HAZTURK based on HAZUS (hazards us) to be used before, during and after a disaster: a case study for Istanbul, in final report to TUBITAK 104Y254 Award. TUBITAK: Ankara, pp 46–117 (2007)
19. Hancilar, U., et al.: ELER software-a new tool for urban earthquake loss assessment. Nat. Hazards Earth Syst. Sci. **10**(12), 2677 (2010)
20. Erdik, M., et al.: Development of rapid earthquake loss assessment methodologies for Euro-Med region. In: Proceedings of 14th World Conference on Earthquake Engineering (2008)

21. Kircher, C.A., Whitman, R.V., Holmes, W.T.: HAZUS earthquake loss estimation methods. Nat. Hazards Rev. **7**(2), 45–59 (2006)
22. Elnashai, A.S., et al.: Overview and applications of Maeviz-Hazturk 2007. J. Earthquake Eng. **12**(S2), 100–108 (2008)
23. Goodchild, M.F.: Citizens as sensors: the world of volunteered geography. GeoJournal **69**(4), 211–221 (2007)
24. Wald, D.J., et al., *USGS "Did You Feel It?" Internet-based macroseismic intensity maps.* 2012, 2012. 54(6).
25. Clarke, K.C.: Getting started with geographic information systems, vol. 3. Prentice Hall Upper Saddle River, NJ (1997)
26. Hotosm: Disaster mapping projects (2017) [01.06.2017]. Available from: https://www.hotosm.org/projects/disaster-mapping
27. Meier, P., Werby, O.: Ushahidi in Haiti: the use of crisis mapping during the 2009 earthquake in Haiti. In: Proceedings of the IADIS International Conference on ICT, Society and Human Beings 2011, Proceedings of the IADIS International Conference e-Democracy, Equity and Social Justice 2011, Part of the IADIS, MCCSIS 2011 (2011)
28. Hirata, E., et al.: Flooding and inundation collaborative mapping—use of the Crowdmap/Ushahidi platform in the city of Sao Paulo, Brazil. J Flood Risk Manag. (2015)
29. Sakaki, T., Okazaki, M., Matsuo, Y.: Earthquake shakes Twitter users: real-time event detection by social sensors. In: Proceedings of the 19th International Conference on World wide web. ACM, Raleigh, pp 851–860
30. Goodchild, M.F.: Citizens as voluntary sensors: spatial data infrastructure in the world of Web 2.0. Int. J. Spat. Data Infrastruct. Res. (2007)
31. Acar, A., Muraki, Y.: Twitter for crisis communication: Lessons learned from Japan's tsunami disaster. Int. J. Web Based Communities **7**(3), 392–402 (2011)
32. Iwanaga, I.S.M., et al.: Building an earthquake evacuation ontology from twitter. In: 2011 IEEE International Conference on Granular Computing (GrC) (2011)
33. Wang, Z., Ye, X., Tsou, M.H.: Spatial, temporal, and content analysis of Twitter for wildfire hazards. Nat. Hazards **83**(1), 523–540 (2016)
34. Issa, E., et al.: Understanding the spatio-temporal characteristics of Twitter data with geotagged and non-geotagged content: two case studies with the topic of flu and Ted (movie). Ann. GIS **23**(3), 219–235 (2017)
35. Ishino, A., et al.: Extracting transportation information and traffic problems from tweets during a disaster. In: Proceedings of IMMM, pp 91–96 (2012)
36. Statista: Most famous social network sites worldwide as of September 2017, ranked by number of active users (in millions). 2017 [cited 2017 02.11.2017]. Available from: https://www.statista.com/statistics/272014/global-social-networks-ranked-by-number-of-users/
37. Hecht, B., Shekhar, S.: From GPS and Google Maps to spatial computing. Coursera Inc. (2014)
38. Turner, A.: Introduction to neogeography (2006). O'Reilly, p. 54
39. Parker, C.J.: A framework of neogeography. In: The Fundamentals of Human Factors Design for Volunteered Geographic Information. Springer, Berlin, pp. 11–22
40. Goodchild, M.: Neogeography and the nature of geographic expertise. J. Locat. Based Serv. **3**(2), 82–96 (2009)
41. Elwood, S., Goodchild, M.F., Sui, D.Z.: Researching Volunteered Geographic Information: Spatial Data, Geographic Research, and New Social Practice. Ann. Assoc. Am. Geogr. **102**(3), 571–590 (2012)
42. Gulnerman, A.G., Gengec, N.E., Karaman, H.: Review of public Tweets over Turkey within a pre-determined time. ISPRS-International Archives of the Photogrammetry, Remote Sensing and Spatial Information Sciences, pp. 153–159
43. Schroeder, P.: Criteria for the design of a GIS/2. specialists' meeting for NCGIA Initiative 19. GIS and Society, Summer (1996)
44. Anderson, L.T.: Guidelines for preparing urban plans (1995)

45. Sieber, R.: Public participation geographic information systems: a literature review and framework. Ann Assoc Am Geogr **96**(3), 491–507 (2006)
46. Gulnerman, A.G., Karaman, H.: PPGIS case studies comparison and future questioning. In: Proceedings—15th International Conference on Computational Science and Its Applications, ICCSA 2015 (2015)
47. Ball, J.: Towards a methodology for mapping 'regions for sustainability' using PPGIS. Prog. Plann. **58**(2), 81–140 (2002)
48. Muller, J.P., et al.: EU-FP7-IMARS: Analysis of Mars multi-resolution images using auto-coregistration, data mining and crowd source techniques: processed results—a first look. In: International Archives of the Photogrammetry, Remote Sensing and Spatial Information Sciences—ISPRS Archives (2016)
49. Wardlaw, J., Jackson, B.: Rights lab: slavery from space. [cited 2017 18.10.2017]. Available from: https://www.zooniverse.org/projects/ezzjcw/rights-lab-slavery-from-space
50. Wardlaw, J., et al.: Mars in motion [cited 2017 18.10.2017]. Available from: https://www.zooniverse.org/projects/imarsnottingham/mars-in-motion
51. Scistarters: Project Finder (2017) [cited 2017 18.10.2017]. Available from: https://scistarter.com/finder
52. Meier, P.: Crisis mapping in action: how open source software and global volunteer networks are changing the world, one map at a time. J. Map Geogr. Libr. **8**(2), 89–100 (2012)
53. Zooniverse: People—powered research 2017 [cited 2017 18.10.2017]. Available from: https://www.zooniverse.org/
54. Scistarters: Science we can do together (2017). Available from: https://scistarter.com/
55. Ushahidi: Read the crowd (2017) [cited 2017 18.10.2017]. Available from: https://www.ushahidi.com/
56. OSM: Open street map about (2016) [cited 2015 25 November]. Available from: https://www.openstreetmap.org/about
57. Valenzuela, S., Arriagada, A., Scherman, A.: The social media basis of youth protest behavior: the case of Chile. J Commun **62**(2), 299–314 (2012)
58. Tufekci, Z., Wilson, C.: Social media and the decision to participate in political protest: observations from Tahrir Square. J. Commun. **62**(2), 363–379 (2012)
59. Lim, M.: Clicks, cabs, and coffee houses: social media and oppositional movements in Egypt, 2004–2011. J. Commun. **62**(2), 231–248 (2012)
60. Haciyakupoglu, G., Zhang, W.: Social media and trust during the Gezi protests in Turkey. J. Comput.-Med. Commun. **20**(4), 450–466 (2015)
61. Tulloch, D.L.: Is VGI participation? From vernal pools to video games. GeoJournal **72**(3–4), 161–171 (2008)
62. Hall, G.B., et al.: Community-based production of geographic information using open source software and Web 2.0. Int. J Geogr. Inf. Sci. **24**(5), 761–781 (2010)
63. Moffitt, J.: Tweet Metadata Timeline (2017) [cited 2017 26.10.2017]. Available from: https://support.gnip.com/articles/tweet-timeline.html
64. Gross, D., Hanna, J.: Facebook introduces check-in feature (2010) [cited 2017 01.11.2017]. Available from: https://edition.cnn.com/2010/TECH/social.media/08/18/facebook.location/index.html
65. Power, R., Robinson, B., Wise, C.: Using Crowd Sourced Content to Help Manage Emergency Events, in Social Media for Government Services. Springer, Berlin, pp. 247–270 (2015)
66. Gelernter, J., Mushegian, N.: Geo-parsing messages from microtext. Trans. GIS **15**(6), 753–773 (2011)
67. Gelernter, J., Wu, G.: High performance mining of social media data. In Proceedings of the 1st Conference of the Extreme Science and Engineering Discovery Environment: Bridging from the eXtreme to the Campus and beyond. ACM (2012)
68. Gelernter, J., Balaji, S.: An algorithm for local geoparsing of microtext. GeoInformatica **17**(4), 635–667 (2013)
69. Leetaru, K., et al.: Mapping the global Twitter heartbeat: the geography of Twitter (2013)

70. Landwehr, P.M., et al.: Using tweets to support disaster planning, warning and response. Saf. Sci. **90**, 33–47 (2016)
71. Utani, A., Mizumoto, T., Okumura, T.: How geeks responded to a catastrophic disaster of a high-tech country: rapid development of counter-disaster systems for the great east Japan earthquake of March 2011. In: Proceedings of the Special Workshop on Internet and Disasters. ACM (2011)
72. Kudou, T., *mecab*. GitHub repository (2013)
73. Ao, J., Zhang, P., Cao, Y.: Estimating the locations of emergency events from Twitter streams. Procedia Comput. Sci. **31**, 731–739 (2014)
74. Yin, J., et al.: Using social media to enhance emergency situation awareness. IEEE Intell. Syst. **27**(6), 52–59 (2012)
75. Castillo, C., Mendoza, M., Poblete, B.: Predicting information credibility in time-sensitive social media. Internet Res. **23**(5), 560–588 (2013)
76. Poorazizi, M.E., Hunter, A.J., Steiniger, S.: A volunteered geographic information framework to enable bottom-up disaster management platforms. ISPRS Int. J. Geo-Inf. **4**(3), 1389–1422 (2015)
77. Crooks, A., et al.: # Earthquake: Twitter as a distributed sensor system. Trans. GIS **17**(1), 124–147 (2013)
78. Rosser, J.F., Leibovici, D.G., Jackson, M.J.: Rapid flood inundation mapping using social media, remote sensing and topographic data. Nat. Hazards **87**(1), 103–120 (2017)
79. Lin, Y.R., Margolin, D.: The ripple of fear, sympathy and solidarity during the Boston bombings. EPJ Data Sci. **3**(1), 1–28 (2014)
80. Terpstra, T., et al.: Towards a Realtime Twitter Analysis During Crises for Operational Crisis Management. Simon Fraser University (2012)
81. Hara, Y.: Behaviour analysis using tweet data and geo-tag data in a natural disaster. Transp. Res. Procedia **11**, 399–412 (2015)
82. Statista: Most popular social networks worldwide as of April 2018, ranked by number of active users (in millions). 2018 [cited 2018 03.05.2018]. Available from: https://www.statista.com/statistics/272014/global-social-networks-ranked-by-number-of-users/
83. Statista: Active social network penetration in selected countries as of January 2018 (2018). Available from: https://www.statista.com/statistics/282846/regular-social-networking-usage-penetration-worldwide-by-country/
84. Statista: Global digital population as of April 2018 (in millions) (2018) [cited 2018 03.05.2018]. Available from: https://www.statista.com/statistics/617136/digital-population-worldwide/
85. Social Media Data Stewardship: Social Media Research Toolkit (2017) [cited 2017 22.10.2017]. Available from: https://socialmediadata.org/social-media-research-toolkit/
86. Gentry, J.: R Based Twitter Client (2016)
87. Cribbin, T.B., Julie, Brooker, P., Basnayake, H.: The Chorus Project Tweet Catcher (2015)
88. KNIME: KNIME Twitter Nodes (2014)
89. Carto: Twitter Data Analysis With a Pulse (2016). Available from: https://carto.com/connectors/twitter-maps/
90. Social Media Research Foundation. NodeXL (2017) [cited 2017 22.10.2017]. Available from: https://www.smrfoundation.org/nodexl/
91. Gengec, N.E.: Geo Tweets Downloader (2015). github. p. Spatial Tweets Downloader
92. Yamamoto, Y.: Twitter4j (2015)
93. Twitter: REST APIs (2015)
94. Twitter: The Streaming APIs (2015)
95. PostgreSQL: Homepage (2017). Available from: https://www.postgresql.org/
96. PostGIS: About PostGIS (2017). Available from: https://postgis.net/
97. Bruns, A., Liang, Y.E.: Tools and methods for capturing Twitter data during natural disasters. First Monday **17**(4) (2012)
98. Zi, C., et al.: Who is tweeting on Twitter: human, bot, or cyborg? In: Proceedings of the 26th Annual Computer Security Applications Conference 978-1-4503-0133-6. ACM, Austin, pp 21–30 (2010)

99. R Core Team: R: A language and environment for statistical computing. R Foundation for Statistical Computing, Vienna, Austria. (2013). http://www.R-project.org/
100. RStudio: Take control of your R code (2016) [cited 2017 06.11.2017]. Available from: https://www.rstudio.com/products/rstudio/
101. R Special Interest Group on Databases. Kirill Müller, H.W.: DBI: R Database Interface (2016)
102. Conway, J., et al.: RPostgreSQL : R interface to the PostgreSQL database system (2016)
103. Basille, M., Bucklin, D.: rpostgis: R Interface to a 'PostGIS' Database (2017)
104. Feinerer, I.: tm: Text Mining Package (2015)
105. Aksoy, A., Ozturk, T.: trstop (2016) GitHub
106. Sadi, E.S.: Metin Madenciliği (Text Mining). In: Bilgisayar Kavramları (2014)
107. Cakir, U., Mehmet, Güldamlasioglu, S.: Text mining analysis in turkish language using big data tools. In: 2016 IEEE 40th Annual Computer Software and Applications Conference (2016)
108. Hearst, M.A.: Untangling text data mining. In: Proceedings of the 37th annual meeting of the Association for Computational Linguistics on Computational Linguistics. 1999. Association for Computational Linguistics
109. Hearst, M.: What is text mining. SIMS, UC Berkeley (2003)
110. Bouchet-Valat, M.: SanowballC: Snowball stemmers based on the C libstemmer UTF-8 library (2014)
111. Evren, K.C.: SnowballC (2007)
112. Stanford NLP Group: Stemming and lemmatization (2008) [cited 2017]. Available from: https://nlp.stanford.edu/IR-book/html/htmledition/stemming-and-lemmatization-1.html
113. Feinerer, I.: Introduction to the tm Package Text Mining in R (2017)
114. ESRI: Optimized Hot Spot Analysis (2016). Available from: https://desktop.arcgis.com/en/arcmap/10.3/tools/spatial-statistics-toolbox/optimized-hot-spot-analysis.htm
115. Cubukcu, K.M.: Planlamada ve coğrafyada temel istatistik ve mekansal istatistik (2015) Nobel
116. ESRI: Hot spot analysis (Getis-Ord Gi*) (2016). Available from: https://desktop.arcgis.com/en/arcmap/10.3/tools/spatial-statistics-toolbox/hot-spot-analysis.htm
117. Red Hen Systems, I.: MapCalcTM Learner-Academic software
118. Hürriyet: Dakika dakika darbe girişimi - 15–16 Temmuz 2016 (2016). Available from: https://www.hurriyet.com.tr/dakika-dakika-darbe-girisimi-15-16-temmuz-2016-40149409
119. NTV: 15 Temmuz gecesi ve sonrasında neler yaşandı (2016). Available from: https://www.ntv.com.tr/galeri/turkiye/15-temmuz-gecesi-ve-sonrasinda-neler-yasandi,fBIiAr1pu0WkrhUBXdZykA/ClgQwqBRtEaLnVzxTJdIXw
120. Gulnerman, A.G., Karaman, H.: Social media spatial monitor of coup attempt in the republic of Turkey. In: 2017 17th International Conference on Computational Science and Its Applications (ICCSA). IEEE (2017)
121. Boccardo, P.: New perspectives in emergency mapping. Eur J Remote Sens **46**(1), 571–582 (2013)
122. Khorram, Y.: As Sandy pounded NYC, fire department worker was a Twitter lifeline (2012) [cited 2017 06.11.2017]. Available from: https://edition.cnn.com/2012/11/01/tech/social-media/twitter-fdny/
123. Lee, K., et al.: Spatio-temporal provenance: Identifying location information from unstructured text. In: 2013 IEEE International Conference on Pervasive Computing and Communications Workshops (PERCOM Workshops). IEEE (2013)
124. Mocnik, F.B., Mobasheri, A., Zipf, A.: Open source data mining infrastructure for exploring and analysing OpenStreetMap. Open Geospat Data, Softw Standards **3**(1), 7 (2018)
125. Fonte, C., et al.: VGI quality control. ISPRS Geospatial week 2015, pp 317–324 (2015)
126. Haklay, M.: How good is volunteered geographical information? A comparative study of OpenStreetMap and ordnance survey datasets. Environ. Plann. B: Plann. Des. **37**(4), 682–703 (2010)
127. Abbasi, M.-A., Liu, H.: Measuring user credibility in social media. In: International Conference on Social Computing, Behavioral-Cultural Modeling, and Prediction. Springer, Berlin (2013)

128. Lenormand, M., et al.: Tweets on the road. PloS one 9(8), e105407 (2014)
129. D'Andrea, E., et al.: Real-time detection of traffic from twitter stream analysis. IEEE Trans. Intell. Transp. Syst. **16**(4), 2269–2283 (2015)
130. Hasby, M., Khodra, M.L.: Optimal path finding based on traffic information extraction from Twitter. In: 2013 International Conference on ICT for Smart Society (ICISS). IEEE (2013)
131. Bassolas, A., et al.: Touristic site attractiveness seen through Twitter. arXiv preprint arXiv:1601.07741 (2016)

The manufacturer's authorised representative in the EU is Springer Nature Customer Service Centre GmbH, Europaplatz 3, 69115 Heidelberg, Germany. If you have any concerns regarding our products, please contact ProductSafety@springernature.com

Printed and bound by CPI Group (UK) Ltd, Croydon, CR0 4YY

30/09/2026

02225515-0013